Microbial Diversity
in Time and Space

Microbial Diversity in Time and Space

Edited by

R. R. Colwell

University of Maryland
College Park, Maryland

Usio Simidu

Tokyo University of Agriculture
Tokyo, Japan

and

Kouichi Ohwada

University of Tokyo
Tokyo, Japan

Plenum Press • New York and London

Library of Congress Cataloging-in-Publication Data

International Symposium on the Microbial Diversity in Time and Space
(1994: Tokyo, Japan)
    Microbial diversity in time and space / edited by R.R. Colwell,
Usio Simidu, Kouichi Ohwada.
        p.   cm.
    "Proceedings of the International Symposium on the Microbial
Diversity in Time and Space, held October 24-26, 1994, in Tokyo,
Japan"--T.p. verso.
    Includes bibliographical references and index.
    ISBN 0-306-45194-8
    1. Bacterial diversity--Congresses.   I. Colwell, Rita R., 1934-
    II. Simidu, Usio.   III. Ohwada, Kouichi.
    QR73.I57   1994
    576--dc20
                                                        96-13455
                                                            CIP

Proceedings of the International Symposium on the Microbial Diversity in Time and Space,
held October 24–26, 1994, in Tokyo, Japan

ISBN 0-306-45194-8

© 1996 Plenum Press, New York
A Division of Plenum Publishing Corporation
233 Spring Street, New York, N. Y. 10013

Printed in the United States of America

PREFACE

The symposium, "Microbial Diversity in Time and Space," was held in the Sanjo Conference Hall, University of Tokyo, Tokyo, Japan, October 24-26, 1994. The symposium was organized under the auspices of the Japanese Society of Microbial Ecology and co-sponsored by the International Union of Biological Sciences (IUBS), International Union of Microbiological Societies (IUMS), International Committee on Microbial Ecology (ICOME), and the Japanese Society of Ecology.

The aim of the symposium was to stress the importance of the global role of microorganisms in developing and maintaining biodiversity. Twenty-four speakers from seven countries presented papers in the symposium and in the workshop, "Microbial Diversity and Cycling of Bioelements," that followed the symposium. Papers presented at the symposium are published in this proceedings. Discussions of the workshop, which were energetic and enthusiastic, are also summarized in this proceedings.

The symposium provided an opportunity to address the role of microorganisms in global cycles and as the basic support of biodiversity on the planet. Previously unrecognized as both contributing to and sustaining biodiversity, microorganisms are now considered to be primary elements of, and a driving force in, biodiversity.

Financial support was provided for the symposium by the CIBA GEIGY Foundation for the Promotion of Science, Naito Foundation, and the Uchida Foundation of the Ocean Research Institute, University of Tokyo. Support from these foundations is gratefully acknowledged.

CONTENTS

MICROBIAL BIODIVERSITY–GLOBAL ASPECTS

Rita R. Colwell

University of Maryland Biotechnology Institute
4321 Hartwick Road, Suite 550
College Park, Maryland 20740

ABSTRACT

The enormous diversity available at the microbial level is beginning to be recognized, but this richness of diversity amongst bacterial and virus species has yet to be cataloged. The number of species of bacteria that have been described is approximately 2,000- 4,000, whereas the estimated total number of bacterial species is approximately 3,000,000, most probably, a significant underestimation. The number of virus species may be in the same range. The role of microorganisms in ecological cycles is just beginning to be more fully understood. For example, viruses have been found to regulate/modulate algal abundance and distribution. Global cycles, including carbon, nitrogen, sulfur, and heavy metal cycles, are known to be mediated or influenced by microorganisms. A unique aspect of the diversity of microorganisms is their morphologyl, ranging from the ultramicroscopic to a recently described bacterial species visible to the naked eye. The shapes of bacteria may be rod, spherical, spiral, triangular, pyramidal, and rectangular, in brief, nearly every possible morphology. Morphology and function of bacteria are influenced by the environment, a demonstration of the inter-relationship of diversity and ecological processes. Inventorying and cataloging microbial diversity is a daunting task, requiring ingenuity and creativity, but offering substantial economic and social reward.

INTRODUCTION

The unit of diversity is the species, which may be defined as: "a group of related organisms that is distinguished from similar groups by a constellation of significant genotypic, phenotypic, and ecological characteristics."

In effect, the microbial species is, at the same time, a naturally-occurring group of similar organisms and represents a taxon defined by collective properties of the group sufficiently to distinguish the species from closely related groups (Tiedje, 1992). The definition of the microbial species includes genotypic traits: structure and composition of

Microbial Diversity in Time and Space, edited by Colwell et al.
Plenum Press, New York, 1996

1

nucleic acids (DNA and RNA); phenotypic traits: morphology (cell size range, shape, composition, and appendages); development (multicellular or dormant stages); metabolism (utilization of elements and energy sources, enzyme structure and regulation); physiology (physiological versatility); and ecological traits: relationship to biotic associations (commensal, symbiotic, and prey-parasite-predator); dependence on and susceptibility to the influence of abiotic factors (pH, temperature, pressure, salinity, illumination, water potential, solute concentration, etc.). The polyphasic approach i.e., including genetic, phenotypic, and habitat/community analysis data (Colwell, 1970a, 1970b) for microbial diversity measurement has been recommended.

Biodiversity is the property of living systems, that is, being distinct, different, unlike, is the very nature of life. Biological diversity or biodiversity can be defined as the characteristic of groups or classes of living entities to be varied. The definition of diversity, given in a standard dictionary, is the condition of being diverse — different, unlike. Thus, each class or entity, e.g., gene, cell, individual, species, community, or ecosystem, has more than one kind. Diversity is a fundamental property of every living system. Because biological systems are hierarchical, diversity manifests itself at every level of biological hierarchy, from molecules to ecosystems (Solbrig, 1991).

Diversity of a system can be divided into units which may differ in the number of entities ("richness" or "abundance") or the components may differ in relative abundance or importance of the entities within them. Thus, diversity can be measured in a variety of ways. One is simply to enumerate the number of distinct units (richness). On the other hand, one can rank species in their order of importance (importance curves) (Whittaker, 1969). A measurement that can be considered more precise is one that considers relative abundance of each type being enumerated, an example being the Shannon-Weaver index of diversity. Because the number of biological entities comprising the global ecosystem is enormous, and most biological species have not yet been described and classified, knowledge of biodiversity is sorely deficient, especially at the microbial level.

Diversity measurements of natural microbial populations were developed mainly to demonstrate effects of perturbations of an econiche. The parameters employed generally include microbial biomass, enumeration of colonies, colonial differentiation, microscopic cell counts, enumeration of species and genera, as well as microbial profiles determined by cluster analysis and factor analysis. The algorithms that have been used in microbial diversity studies originated, in the main, in the physical sciences, psychology, and in plant and animal ecology. Diversity formulas used in ecology have been presented in detail by Pielou (1975).

A diversity index, even without the discriminating power of factor analysis to elucidate variance in representative samples from large geographical areas, can be helpful in developing succession in a limited econiche. The process of succession is one of the binding concepts of ecology, and a salient notion for microbial ecology, where constantly changing environments, viz., temperature, nutrients, ion concentration, pH, and oxygen concentration, create pressure for change in microbial populations. Microbial succession involves changes in species composition and abundance.

Since species diversity comprises species abundance, evenness, equitability, and richness, simple averages and percentages are much too limited for describing microbial diversity. Isolating and identifying individual organisms in environmental samples, such as a liter of water or 1,000 g of soil comprise the classic approach to assessing species density, taking into account the spatial component. From such data, estimates of microbial populations of the entire body of water are made. In reality, measurement of diversity, the variety of flora or fauna in an econiche, requires knowledge of the richness (number of species per sample), evenness (relative abundance of species, equitability), and diversity (total number of organisms per sample). Diversity indices used in plant and animal ecology have been applied to microbiology (Russek–Cohen and Jacobs, 1995).

RATIONALE FOR PROMOTING MICROBIAL DIVERSITY RESEARCH AND EDUCATION

The importance of microbial diversity can be realized by determining the extent of microbial diversity, assessing the importance of diversity to the stability of the biosphere, monitoring and predicting environmental change using microorganisms as indicators of environmental health and as indicator organisms for pollutants, mineral resources, anthropogenic contamination, and by expanding biotic resources through food processing, waste removal, antibiotics discovery, and biotechnology.

Edward Kellenberger (1994), in a thoughtful essay, defined "genetic ecology" as a new interdisciplinary science, namely an extension of molecular genetics to studies of viability (dormancy), gene expression, and gene movements in natural environments such as soils, aquifers, the oceans, and digestive tracts. "Genetic ecology" — specifically includes movements of genetic material within and between species, and even kingdoms, in natural environments. Gene transfer between bacteria in the laboratory once was assumed to be confined to very few species and in the natural environment these exchanges were believed not to occur. This assumption has proven false. Horizontal gene transfer is actually much more frequent, extrapolating from laboratory experiments. Thus, the study of free DNA, plasmids, mobile elements, etc., in natural environments must, of necessity, be a part of the analysis of microbial diversity. Transduction of genes by viruses in natural environments is clearly a potentially major contribution to genetic variability. Therefore, considering the continuous flow of genetic material, the genetic stability of species is quite remarkable. Areas of fruitful study in genetic ecology are microbial colonization and symbiosis, e.g., gut of animals, lichens, etc. In fact, the colonization of roots of plants by rhizobia may have many features in common with colonization of the gut of animals. Transformation via free DNA and conjugation in the natural environment may be significant. For example, since DNA absorbs strongly to minerals, it is 100 fold more resistant to DNAases.

Interspecies transfer is an exciting area of research. Gene exchange between a coliphage and a fungus has been reported. We must remember that bacterial transformation of pneumococci in 1928 was an idea ridiculed at the time! Even dormant (viable, but nonculturable) cells may transfer DNA to other, actively growing cells and vice versa, possibly the microbial equivalent of polyploidy and geographic speciation giving rise to new species in higher organisms.

In marine samples, typically less than one per cent of the bacteria visualized microscopically are culturable by current techniques. The conclusion is that assessment of overall bacterial diversity by examining only the culturable portion of a population is inadequate (Colwell and Hill, 1993). To study the biodiversity of the nonculturable portion of a marine population, it is necessary to use molecular techniques, such as gene probing and polymerase chain reaction (PCR) amplification of 16S rRNA genes (Giovannoni et al., 1990; Knight et al., 1990, 1991). The study of genetic diversity of marine bacterial populations has been greatly advanced by sequence analysis of PCR-amplified 16S rRNA genes. This approach has been used to demonstrate the presence of a novel microbial group in the Sargasso Sea, supporting the view that microbial ecosystems contain new uncultivated species (Giovannoni et al., 1990). This has been further indicated by the remarkable finding that archaebacteria are present in large numbers in pelagic marine bacterial populations (Fuhrman, 1992, 1993).

ESTIMATING MICROBIAL BIODIVERSITY

An early attempt to solve a microbial ecology problem using diversity measurements of samples obtained in the field was made, whereby the bacterial diversity of a pristine salt marsh sediment was compared with the diversity found in an oil field marsh sediment. In this case, the diversity index of Wilhum and Dorris (1969) was used to measure bacterial diversity according to colony type. The problem with this approach, from a mathematical point of view, is lack of correlation between phenome and genome, which can be overcome by the more recently developed method of mRNA tagging.

Lelong et al. (1980) compared the taxonomy, ecological profiles, diversity, and average catabolic potential of two experimental phytoplankton blooms. High diversity was associated with oligotrophic conditions and phytoplankton exponential growth phase. Diversity decreased during plankton mortality, with vibrios being predominant after onset of the mortality phase. Bacterial diversity preceded the appearance of the phytoplankton blooms and continued at a high rate. Stable states were reflected by high diversity, whereas decrease in diversity was associated with sudden perturbations, e.g., phytoplankton death. Phytoplankton, decline is followed by a burst of zooplankton biomass. Since Simidu and colleagues have shown that pseudomonads are associated with phytoplankton and vibrios with zooplankton blooms, the decreased diversity, i.e., dominance in vibrios can be related to association of vibrios with components of the zooplankton, namely copepods.

Information about bacterial communities has been assembled by painstakingly aggregating data for individual members. Spatial and temporal distribution of organisms have been difficult to measure, but molecular genetic methods for examining bacterial communities *in toto* now make this an easier task. In fact, significant advances have been made in assessing bacterial and fungal biodiversity, but the methods employed have technical or interpretation limitations. There is not any single method that can provide an unequivocal estimate of bacterial or fungal diversity. Culture methods, or enumeration by direct microscopy, have been the classical approach, but the data obtained using these methods are incomplete and may not be very important *in situ*. The indirect methods that are used to correlate diversity with processes between different sites can be more helpful in this regard.

Methods for determining bacterial biodiversity include chemosystematic determinations of taxon-specific cellular constituents, such as fatty acids, sterols, secondary metabolites, protein structures, etc. Significant advances have been made recently using nucleic acid-based methods, such as whole community nucleic acid hybridization, community DNA reassociation kinetics, cloning and sequencing of polymerase chain reaction (PCR)-amplified 16S rNA gene sequences from community DNA samples, as well as nucleic acid, function-based methods, such as hybridization of probes to genes (which measures potential function) or mRNA (expressed function) specific to given enzymatic activities. The combination of traditional methods of culture and microscopy with molecular identifications can be valuable in measuring diversity of microbial populations.

Hofle (1992) examined the composition and dynamics of whole marine bacterioplankton communities by analysis of low molecular weight RNA (LMW RNA) isolated directly from water samples, finding that, in natural bacterial communities, enough variability exists in the pool of LMW RNA (primarily 5S rRNA and tRNAs) that simply resolving the mixtures of LMW RNA in these pools, using polyacrilamide electrophoresis, yielded banding patterns characterizing each community. That is, bacterial communities contain so much genetic diversity that size analysis of products of even a few genes yielded enough heterogeneity to provide community-specific patterns.

Evaluating spatial and temporal variability of marine communities requires examination of a large number of samples. Practical analytical methods, if to be used in field studies, therefore must be simple, rapid, and reproducible. Direct comparison of DNA recovered from water samples, using the randomly amplified polymorphic DNA (RAPD) method of Williams et al. (1990) to generate DNA fingerprints of bacterial communities characteristic of such community has been employed. The fingerprints provide a means of comparing whole bacterial communities in different water samples, allowing elucidation of spatial and temporal variation of bacterial communities in Chesapeake Bay (Straube et al., 1996).

To illustrate, a water sample is collected at a specific station and inoculated on LB plates. Colonies are isolated and used to prepare overnight broth cultures from which DNA is prepared (Ausabel et al. 1994). Community DNA is recovered from water samples (Somerville et al., 1989) and, briefly, a particulate fraction is concentrated from each water sample by filtration through a filter cartridge of 0.22 μm pore size.

Targeted and nontargeted primers are prepared. In studies carried out in Chesapeake Bay, a targeted primer was an 18-mer, (5' GTA TTA CCG CGG CTG CTG 3'), designed to hybridize to a highly conserved region (bases 521-539, using the *Escherichia coli* numbering convention) of the 16S rRNA gene, and to direct polymerization toward the 5' end of the gene and into the upstream control region. A second primer was a decamer of random sequence, with a 60% G+C content.

DNA fingerprints of samples collected from a Chesapeake Bay site was capillary blotted (Sambrook et al., 1989) onto nylon hybridization membranes and prehybridization and hybridization were carried out. An eubacterial probe (Amann et al., 1990, 1991) targeted to bases 338-355 (*Escherichia coli* numbering convention) of the 16S rRNA gene was end-labeled with [gamma-^{32}P]ATP and hybridized to the membrane and autoradiographs of the gel were prepared.

The RAPD method has been used by several investigators to distinguish among strains of bacteria (Cancilla et al., 1992) and fungi (Kersulyte et al., 1992), and to examine sequence divergence among derivatives of *Escherichia coli* K-12 (Brikun, 1994). The utility of RPD in differentiating among closely related strains is due to the fact that small changes in the base sequence of the genome can affect where primers hybridize, thus altering or eliminating one or more of the bands in the DNA fingerprint.

Semirandomly primed PCR produced a relatively simple, but unique fingerprint from estuarine bacterial isolates and the community fingerprints generated by this method were found to be suitably complex to yield useful comparisons. Community fingerprints generated from water samples collected in the Chesapeake Bay showed strong stratification in the Bay mid-spring. The surface waters of Chesapeake Bay are dominated by input from the Susquehanna River (Boicourt, 1992). Thus, at mid-spring, if the temperature, salinity and nutrient composition of surface waters are similar mid-Bay, the species composition of the bacterial communities in the water samples can be predicted to be similar. Finding a spatially large microbial community emphasizes the utility of this method in determining an appropriate scale for microbial ecological studies in the future (Straube et al., 1996).

Different water samples collected at a single station, but at several times during the year, revealed the presence of different bacterial communities. This finding is in agreement with variation in temperature, salinity and nutrient conditions that occurred in the area from which the samples were taken during the annual seasonal cycle (Straube et al., manuscript in preparation).

SOURCE OF SPECIES DIVERSITY

Because the nucleic acid molecule (DNA and RNA) is the information carrier, it has the remarkable capacity to maintain its physical-chemical integrity with almost infinite variability possible in the nucleotide order and, therefore, differentiation and mutation create species diversity. It has been hypothesized by some investigators that no new genetic information is created, instead, existing information is partitioned into discrete entities. Even though it appears that species operate independently, when communities and ecosystems are formed, the obvious and significant factor is that all species are functionally linked with other species, in order to form communities and ecosystems.

It is important to distinguish processes that originate diversity, those that maintain diversity, and those that reduce diversity. Mutation and genetic recombination affect individual and population diversity. Selection encompasses a variety of processes that eliminate variation. New diversity is constantly injected into biological systems — from cells to ecosystems (via mutation, recombination, and related events). On the other hand, diversity is constantly being eliminated by selection. Marine environments, both temperate and tropical, represent the least known of all ecosystems and offer a good source of material to study these processes.

In contrast to what had been predicted a century ago, the oceans have been shown to have the highest diversity of animal and plant phyla. Until recently, the deep sea was believed to be azoic, yet an abundant biota, with more than 800 known species and more than 100 families in a dozen phyla have been reported (Grassle, 1989, 1991). The ocean hydrothermal vents, the sulfide chimneys called "black smokers," contain at least 16 new families of invertebrates that were completely unknown several years ago (Grassle, 1989, 1991). Picoplankton were discovered in cell sizes of 1-4 microns that have yet to be cultured. Because the marine picoplankton are little known, the productivity of marine systems probably is underestimated by at least 50%. Lack of appropriate methods of measurement have contributed to this relative ignorance.

BIODIVERSITY AND ECOSYSTEM PROCESSES

Along with the increasing concern of loss of species diversity and reduction in genetic diversity of crops and wild species, loss of ecosystem structure and function may be occurring on a global scale. This concern arises mainly because of the rapid rate of destruction of tropical landscapes, especially forested landscapes. Because it is not clear how diversity of genes, genotypes, species, and communities affect ecosystem function, it is difficult to assess the global effect of diversity loss on such a massive scale. One approach, at the microbial level, may be to examine microbially mediated processes in soil, as a potential paradigm for larger systems.

A very important and dynamic focal point for a large proportion of ecosystem processes, both natural and anthropogenically-influenced ecosystems, is represented by soil. The dynamic processes vital to global sustainability include turnover of nutrients, uptake of nutrients by plants, soil fertility, formation of soil organic matter, fixation of nitrogen, production of methane, production of CO_2, development of the soil matrix itself, and production of organic acids involved in weathering of rocks. These comprise dynamic processes important in sustainable life. Interestingly, the major global storage reservoirs for carbon are soils, the carbon being in the form of organic matter. Some estimates suggest that *ca.* 1,500 x 10 x 15 g C is stored in soil. The communities of microorganisms, fungi, and invertebrates comprise food webs in soils and are responsible for the cycling of carbon and nitrogen which involve several decomposition steps to yield compounds available for growth of plants and, at the same time, contributing to the rate of production and consumption of CO_2, methane, and nitrogen.

Global climate change can alter the balance of carbon fluxes. The annual flux of CO_2 returning to the atmosphere that is derived from decomposition and other processes taking place in soil has been calculated to be *ca.* 68 x 10 x 15 g C/yr. These changes include land use, as well as climate change itself, involved in changes in spatial and temporal distribution of ecosystem resources. The impact is not only on the diversity of vegetation patterns, but on that of the soils and soil biotaxa involved in decomposition and release of greenhouse gases. Unfortunately, it is not clear just how organisms in soil function within soil systems. Soil biodiversity, structure and interactions of soil organism communities and involvement of soil microbiology in ecological processes need to be better understood.

Moore and Deruiter (1991) have provided information which shows that biodiversity in soils is structured into food chains and webs that are very important determinants of effective functioning of ecosystems. Soil food webs also play a role in providing food for the various above-ground food chains, including small mammals. Loss of biodiversity can diminish ecosystem processes functioning. An obvious example of anthropogenic activities affecting soil biodiversity is addition of fertilizers, which increases plant productivity, but masks or mutes the importance of soil biota in providing nutrients for the plants. Despite the various effects on soil, biodiversity of soil, in general, is believed to be greater than above ground diversity.

A serious problem is that soil microorganisms are difficult to identify to species. It has been suggested by participants in a workshop held at the National History Museum, London, England in 1994 that, because soil is an opaque medium and *in situ* identification of most organisms is presently impractical, it is difficult to identify species in a soil habitat. Furthermore, organisms in soil represent a number of phyla, from microorganisms to earthworms, making it difficult to assess interactions and ecological roles. Also organisms in soil range in size from viruses to earthworms, and the variation in morphology within a given taxon can correlate with life cycle. It has long been known to microbiologists that methods for extracting many of the microorganisms — including fungi and mesofauna — from soils have not been worked out effectively. The techniques for culturing certainly are even less well developed, creating serious problems for identification and enumeration of these microorganisms.

Systematists expert in identifying organisms in soil and involved in critical soil ecosystem functions are few and far between — in fact, a result of the lack of emphasis that soil taxa have received is the lack of appreciation of them as critical components of soil ecosystems. And, finally, the time and spatial scales of the soil habitat, ranging from soil particles to entire landscapes will vary with the taxa present in soil food webs and these will change with changes in the physical/chemical environment. Thus, the cause and effect relationship between loss of species and impact on terrestrial and global ecosystem processes is yet to be fully understood.

Sampling and identification methods, unfortunately, are taxon-specific, and many techniques are in their developmental infancy. There is no single extraction or collection method that will, by itself, extract quantitatively, or collect all microorganisms, or even a single phylum of organisms in a given soil or marine sediment sample. Sampling and extraction methods are chosen based on the experimental question asked, taking into account parameters of the soil habitat and the limitations of the methods used. There have been advances in methodology, but these have been relatively limited. Species distribution, abundance, population structure, and ecological roles remain to be fully elucidated. Taxonomic information for soil biota has been summarized by several inves-tigators (Groombridge, 1992; Hawksworth and Ritchie, 1993; Systematics Agenda 2000, 1991, 1994).

In addition to studying existing microbial biodiversity, it is important to gauge the effects that anthropogenic forces are exerting on this biodiversity. Pollution may be having major, undetected effects on microbial diversity in the marine environment. Long-term monitoring programs are necessary to detect changes of this type. An example of the type of program required is the Long-term Ecosystem Observatory (LEO-2500) that has been established on the Continental Slope off the coast of New Jersey under the direction of Fred Grassle of the Institute of Marine and Coastal Sciences at Rutgers. This site, where water depths are approximately 2,500 meters, has been impacted by sewage sludge dumping since 1986. It has been shown that sewage affects the benthic environment (Hill et al., 1992a, 1993; Bothner et al., 1992). Furthermore, the site serves as a useful model for assessing the impact of widespread eutrophication of the benthic environment at the level of microbial diversity. Indications are that the sludge dumping has had profound effects on the benthic microbial ecology (Straube et al., 1992). DNA extracted from samples collected at the dumpsite have yielded a "genetic snapshot" of the total microbial population at a particular time, providing an ideal archival record of the biodiversity of the population, and serving as a valuable resource for assessing future changes in microbial diversity. It should be noted that microbes are ideally suited to this type of analysis, which is not possible to the same extent with macrobiota. Studies of this type may, therefore, be crucial in obtaining early warning of changes in biodiversity in the oceanic realm.

The recent discovery that viruses are far more numerous in marine systems than was previously realized (Bergh et al., 1989; Proctor and Fuhrman, 1990) has interesting implications for microbial diversity. Viruses may be an important factor in controlling bacterial populations, and it has been suggested that viral infections may be a source of selective pressure contributing to the high bacterial biodiversity in ocean systems (Giovannoni et al., 1990). Seasonal changes in virus populations have been documented for the Chesapeake Bay, suggesting that viruses may be an important trophic factor (Wommack et al., 1992). Preliminary results indicate that viruses may influence bacterial productivity. It has also been shown that viruses can affect phytoplankton primary production. By lysing bacteria and phytoplankton, viruses may divert carbon away from larger bacteriovores and herbivores and consequently return carbon, which would otherwise be utilized at higher trophic levels, to oceanic dissolved organic carbon pools. Viruses may, therefore, be an important factor influencing global carbon budgets, which, in turn, have a major impact on climate change.

In addition to the influence that viruses may have on microbial biodiversity, the biodiversity of virus populations in the marine environment remains to be fully elucidated. A differential filtration procedure that results in bacterium-free virus concentrates, suitable for DNA extraction has been developed (Hill et al., 1992b). This method allows gathering "genetic snapshots" of virus DNA in a particular ecosystem, which then can be compared with time and spatial scale samples to assess changes in virus biodiversity over geographic distance, as well as seasonally.

BIODIVERSITY AND BIOTECHNOLOGY

Biodiversity as a source of innovation in biotechnology has been reviewed by Bull et al. (1992). The approach to biotechnological application can either be directed, with a commercial target in mind, or creativity-driven. Examples include search for actinomycetes [success/application-driven] and alkalophilic microorganisms (creativity-driven). Exploration of extreme environments, e.g., alkaline, hyper-saline, and hyper-thermophilic environments, has been spectacularly successful.

By gaining an understanding of microbial biodiversity, very tangible and direct benefits can accrue. The marine environment, for example, is proving to be a valuable source of novel bioactive compounds with antibacterial, antiviral, and anticancer properties. Free-living bacteria in the sea and bacteria that are symbionts of marine invertebrates are proving to be good sources of useful bioactive compounds. Marine sponges, in particular, which contain diverse communities of bacteria, produce many classes of compounds that are unique to the marine environment.

The filamentous bacteria, the actinomycetes are exceptionally important in biotechnology: they are remarkably versatile metabolically and produce more than 70 percent of the antibiotics currently in medical use. Traditionally, terrestrial actinomycetes have been screened for novel compounds, but marine actinomyocetes have been little investigated, although they have been shown to be ubiquitous in marine sediments. The biodiversity of marine actinomycetes is considerably different from that of terrestrial actinomycetes (Jensen et al., 1991; Takizawa and Hill, 1992; Takizawa et al., 1993). Following on from earlier work (Walker and Colwell, 1975), actinomycetes isolated from the Chesapeake Bay comprise a markedly different species distribution from those found in terrestrial samples (Takizawa et al., 1993). Marine actinomycetes are potentially a very important source of novel bioactive compounds, but a necessary prerequisite for biotechnological exploitation of this group is to gain an understanding of their biodiversity.

CONCLUSIONS

As E. O. Wilson (1988) has stated, "Biological diversity must be treated more seriously as a global resource, to be indexed, used, and above all, preserved. Three circumstances conspire to give this .... unprecedented urgency: ... First, exploding human populations are degrading the environment at an accelerating rate, specially in tropical countries. Second, science is discovering new uses for biological diversity in ways that can relieve both human suffering and environmental destruction. Third, much of the diversity is being irreversibly lost through extinction caused by the destruction of natural habitats ... Overall, we are locked in a race."

The amount of biological diversity globally is suggested to comprise *ca.* 1.4 M living species of all kinds of organisms that have been described. Approximately, 750,000 insect, 41,000 vertebrate and 250,000 plant (vascular plants and bryophytes) species are included in this estimate. The remainder are a complex array of invertebrates, fungi, algae, and microorganisms. If insects are included, Wilson (1988) believes that the absolute number is likely to exceed five million. However, the collections of Terry L. Erwin and associates in the canopy of the Peruvian Amazon rain forest have moved the plausible upper limit to perhaps 30 million species (Erwin, 1982).

The number of genes range from 1,000 in a bacterium and 10,000 in some fungi, to 400,000 or more in many flowering plants and a few animals. (The house mouse (*Mus musculus*)) has about 100,000 genes. Human beings have genetic information closer in amount to the mouse than to salamanders and flowering plants. Clearly, the genetic diversity of the earth is staggering.

Microbiologists are ideally positioned to tie microbial diversity to economic wealth of nations. The task before us is to undertake inventorying, description, cataloguing, and data storage of microorganisms. The magnitude of the undertaking is formidable, but feasible and affordable. For the cost of a fighter jet plane or two, the richness and biological glory of the planet can be realized. A fabulous return for such a minor investment!

If the coral reefs, the floor of the deep sea, and the soil of tropical forests and savannas, are included, somewhere between 5-30 million species may exist on this planet. From island biogeography, the rule of thumb is that a 10-fold increase in area results in the doubling of the number of species. Think of the implication, if this rule holds for microorganisms!

REFERENCES

Amann, R. I., Krumholz, L. and Stahl, D. A. 1990. Fluorescent-oligonucleotide probing of whole cells for determinative, phylogenetic and environmental studies in microbiology. J. Bacteriol. *172*: 762-770.

Amann, R. I., N. Springer, W. Ludwig, H.-D. Görtz, and K.-H. Schleifer. 1991. Identification of *in situ* phylogeny of uncultured bacterial endosymbionts. Nature *351*: 161-164.

Ausabel, F. M., R. Brent, R. E. Kingston, D. D. Moore, J. G. Seidman, J. A. Smith, and K. Struhl. 1994. Current protocols in molecular biology. Greene Publishing Associates and John Wiley & Sons, Inc., NY.

Boicourt, W. C. 1992. Influences of circulation processes on dissolved oxygen in the Chesapeake Bay. p. 7-53. *In*: D. A. Smith, M. Leffler, and G. Mackierman (ed.), Oxygen Dynamics in the Chesapeake Bay: Synthesis of Recent Research, Maryland Sea Grant, College Park, MD.

Bergh, O., K. Y. Boraheim, G. Bratbak, and M. Heidal. 1989. High abundance of viruses found in aquatic environments. Nature. *340*: 467-468.

Bothner, M. H., H. Takada, I. T. Knight, R. T. Hill, C. Butman, J. W. Farrington, R. R. Colwell, and J. F. Grassle. 1992. Sewage contamination in sediments beneath a deep-ocean dumpsite off New York. Marine Environ. Res. *38*: 43-59.

Brikun, I. 1994. Suziedelis, Kestutis: Berg, Douglas E. DNA Sequence Divergence among Derivatives of *Escherichia coli* K-12 Detected by Arbitrary Primer PCR (Random Amplified Polymorphic DNA) Fingerprinting. Jnl. of Bacter. Mar. Vol. 176, No. 6, pp. 1673-1682.

Bull, A. T., M. Goodfellow, and J. H. Slater. 1992. Biodiversity as a source of innovation in biotechnology. Ann. Rev. Microbiol. *46*: 219-252.

Cancilla, M. R., I. B. Powell, A. J. Hiller, and D. E. Davidson. 1992. Rapid genomic fingerprinting of *Lactococcus lactia* strains by arbitrarily primed polymerase chain reaction with ^{32}P and fluorescent labels. Appl. Environ. Microbiol. *58*: 1772-1775.

Colwell, R. R. 1970a. Polyphasic taxonomy of bacteria. Culture Collections of Microorganisms. Poc. International Conference on Culture Collections, Oct 1968, Tokyo: Tokyo Univ. press pp. 421-436.

Colwell, R.R. 19700b. Polyphasic taxonomy of the genus *Vibrio*: numerical taxonomy of *Vibrio cholerae*, *Vibrio parahaemolytions*, and related *Vibrio* species. J. Bacteriol. *104*: 410-433.

Colwell, R. R. and R. Hill. 1993. Microbial diversity. *In*: M. N. A. Peterson (ed). *Diversity of Oceanic Life: An Evaluative Review*, Ocean Policy Institute. Honululu and the Center for Strategic and International Studies. Washington, DC. pp. 100-106.

Erwin, T. L. 1982. Tropical forests: Their richness in *Colsoptera* and other arthropod species. Coleopterist's Bulletin. *36*: 74-75.

Fuhrman, J., K. McCallum, and A. A. Davis. 1992. Novel major archaebacterial group from marine plankton. Nature. *356*: 148-149.

Fuhrman, J., K. McCallum, and A. A. Davis. 1993. Phylogenetic diversity of subsurface marine microbial communities from the Atlantic and Pacific Oceans. Appl. Environ. Microbiol. *59*: 1294-1302.

Giovannoni, S. J., T. B. Britcschgi, C. L. Moyer, and K. G. Field. 1990. Genetic diversity in Sargasso Sea Bacterioplankton. Nature. *345*: 60-63.

Grassle, J. F. 1989. Species diversity in deep-sea communities. Trends Ecol. Evol. *4*: 12-15.

Grassle, J. F. 1991. Deep-sea benthic biodiversity. Bioscience. *41*: 464-469.

Groombridge, B. (ed.). 1992. Global Biodiversity: Status of the Earth's Living Resources. London. Chapman and Hall.

Hawksworth, D. L. and J. M. Ritchie. 1993. *Biodiversity and Biosystematic Priorities: Microorganisms and Invertebrates. Priorities for Biosystematic Research in Support of Biodiversity in Developing Countries: Microorganisms and Invertebrates.* CAB International, Wallingford, Oxon, U.K. 120 pp.

Hill, R. T., I. T. Knight, M. Anikis, W. L. Straube, and R. R. Colwell. 1992a. Benthic distribution of sludge indicated by *Clostridium perfringens* spores at a sewage disposal site off the coast of New Jersey. Amer. Geophys. Union Ocean Science Meeting, New Orleans, LA.

Hill, R. T., K. E. Wommack, and R. R. Colwell. 1992b. Bacterium-bacteriophage interactions in the Chesapeake Bay. 92nd General Meeting of the Amer. Soc. Microbiol., New Orleans, LA.

Hill, R. T., I. T. Knight, M. Anikis, and R. R. Colwell. 1993. Benthic Distribution of Sewage Sludge Indicated by *Clostridium perfringens* at a Deep-Ocean Dump Site. 1993. Appl. Environ. Micriobiol. *59*(1): 47-51.

Höfle, M. G. 1992. Aquatic microbial community structure and dynamics during large-scale release of bacteria as revealed by low-molecular-weight. RNA analysis. Appl. Environ. Microbiol. *58*: 3387-3394.

Jensen, P. R., R. Dwight, and W. Fenical. 1991. Distribution of actinomycetes in near-shore tropical marine sediments. Appl. Environ. Microbiol. *57*: 1102-1108.

Kellenberger, E. 1994. Genetic ecology: A new interdisciplinary science, fundamental for evolution, biodiversity and biosafety evaluations. Experientia. *50*: 429-437.

Kersulyte, A., J. P. Woods, E. J. Keath, W. E. Goldman, and D. E. Berg. 1992. Diversity among clinical isolates of *Histoplasma capsulatum* detected by polymerase chain reaction with arbitrary primers. J. Bacteriol. *174*: 7075-7079.

Knight, I. T. S. Schults, C. W. Kasper, and R. R. Colwell. 1990. Direct detection of *Salmonella* spp. in estuaries by using DNA probe. Appl. Environ. Microbiol. *59*: 1059-1066.

Knight, I. T., J. DiRuggiero, and R. R. Colwell. 1991. Direct detection of enteropathogenic bacteria in estuarine water using nucleic acid probes. Water Sci. Technol. *24*: 261-266.

Lelong, P. P., M. A. Bianchi, and Y. P. Martin. 1980. Planktonic and Bacterial Population Dynamics During Experimental Production of Natural Marine Phytophytoplankton II. Structure and Physiology of Populations and Their Interactions. Canadian Journal of Microb. Vol. 26, pp. 297-307.

Moore, J. C. and P. C. Deruiter. 1991. Temporal and Spatial Heterogeneity of Trophic Interactions Within Belowground Food Webs. Agri. Ecosystems and Environ. Vol 34, pp.371-397.

Pielou, E. C., 1975. *Ecological Diversity*. New York. Wiley.

Proctor, L. M. and J. A. Fuhrman. 1990. Viral mortality of marine bacateria and cyanobacteria. Nature. *343*: 60-62.

Russek-Cohen, E. and D. Jacobs. 1995. Statistics, Biodiversity, and Microorganisms. *In*: (eds.) D. Allsopp, R. R. Colwell, and D. L. Hawksworth. *Microbial Diversity and Ecosystem Function*. CAB International, Wallingford, Oxon, U.K., pp. 305-320.

Sambrook, J., E. F. Fntsch and T. Maniatis. 1989. *Molecular Cloning*. Cold Spring Harbor Press, Cold Spring Harbor, NY.

Solbrig, O. T. (ed.). 1991. *From Genes to Ecosystems: A Research Agenda for Biodiversity*. Cambridge, MA. International Union of Biological Sciences, Paris.

Somerville, C. C., I. T. Knight, W. L. Straube, and R. R. Colwell. 1989. Simple, rapid method for direct isolation of nucleic acids from aquatic environments. Appl. Environ. Microbiol. *55*: 548-554.

Straube, W. L., M. Takizawa, R. T. Hill, and R. R. Colwell. 1992. Response of near-bottom pelagic bacterial community of a deepwater sewage disposal site to deep-sea conditions. Amer. Geophys. Union Ocean Sciences Meeting, New Orleans, LA.

Straube, W. L., R. T. Hill, and R. R. Colwell. 1996. Comparison of aquatic bacterial communities using semirandomly primed PCR. (Submitted)

Systematics Agenda 2000. 1991. Systematics Agenda 2000: Integrating biological diversity and societal needs. Systematic Botany. *16*: 758-761.

Systematics Agenda 2000: Charting the Biosphere - A Global Initiative to Discover, Describe and Classify the World's Species. Technical Report. 1994. Amer. Soc. Plant Taxon., Soc. System Biologists, Willi Hennig Society, and Assoc. System. Coll., 34 pp.

Takizawa, M. and R. T. Hill. 1992. Isolation and ecological studies of actinomycetes in the Chesapeake Bay. 92nd General Meeting of the Amer. Soc. Microbiol., New Orleans, LA.

Takizawa, M., R. T. Hill, and R. R. Colwell. 1993. Isolation and diversity of actinomycetes in the Chesapeake Bay. Appl. Environ. Microbiol. pp. 997-1002.

Tiedje, J. 1992. Microbial diversity: Priorities for research and infrastructure. A Conference Speonsored by the Center for Microbial Ecology and the Bergey's Manual Trust, June 15-18, 1992, Michigan State University, East Lansing, MI.

Walker, J. D. and R. R. Colwell. 1975. Factors affecting enumeration and isolation of actinomycetes from Chesapeake Bay and southeastern Atlantic Ocean sediments. Marine Biol. *30*: 193-201.

Whittaker, R. H. 1969. New concepts of kingdoms of organisms. Science. *163*: 150-160.

Wilhum and Dorris, (1969).

Williams, J. G. K., A. R. Kublik, K, J. Livak, J. A. Rafalski, and S. V. Tingey. 1990. DNA polymorphisms amplified by arbitrary primers are useful as genetic markers. Nuc. Acids Res. *18*: 6531-6535.

Wilson, E. O. 1988. *Biodiversity* (Ed. E. O. Wilson), pp. 1-18. Washington, DC. National Academy Press.

Wommack, K. E., R. T. Hill, M. Kessel, E. Russek-Cohen, and R. R. Colwell. 1992. Distribution of viruses in the Chesapeake Bay. Appl. Environ. Microbiol. *58*: 2965-2970.

2

IMPORTANCE OF COMMUNITY RELATIONSHIPS IN BIODIVERSITY

Hiroya Kawanabe

Lake Biwa Museum
10091 Oroshimo
Kusatsu
Shiga 525, Japan

COMMUNITY RELATIONSHIPS IN BIODIVERSITY

It is well known, even by the general public, that there are many levels of biodiversity, such as the gene, species, higher taxa, and the ecosystem or landscape. All are important and significant components of biodiversity: i.e., the diversity of living materials or of 'creatures' themselves.

There is another kind of biodiversity, however: i.e., the complex, sophisticated interactions among life forms, such as social interactions, competition, cooperation or symbiosis, predation, parasitism, and interrelationships among these interactions. Such a diversity of ecological relationships is not less, but even more, an important part of diversity than simply the visible diversities among 'creatures', as mentioned above (Kawanabe, et al., 1993). Two very 'simple' examples are offered, as follows.

One is competitive cooperation or exploitative mutualism among fishes, first discovered in cichlids of Lake Tanganyika at the end of the 1970s. There are several species of scale-eaters, which pick up scales of other individual fish swimming around them. In the usual case, the success ratio of feeding for each individual scale-eater is rather low, less than 20% on average. If individuals belonging to different species are together and available to attack, however, the success ratio for either individual increases to more than double on average (Fig. 1) (Hori, 1983, 1987, 1990). Individual victims may be much more vulnerable when simultaneously attacked, with two different behaviours of attack. Such kinds of mutualistic situations in fishes were detected also among various benthic animal feeders, piscivores, and algal feeders. (Hori, 1983, 1987; Kohda, 1994; Nakai, 1993; Takamura, 1983; Yuma, 1993, 1994).

Furthermore, in the case of all scale eaters, none of the individuals have a mouth at their terminal end, but, instead the mouth is twisted to the left or to the right side and this feature is genetically determined. It is a simple Mendelian inherited trait. Dextral and sinistral mouths in the same species of the scale eater also allow exploitative mutualism for the scale eating species (Hori, 1990, 1993).

Microbial Diversity in Time and Space, edited by Colwell et al.
Plenum Press, New York, 1996

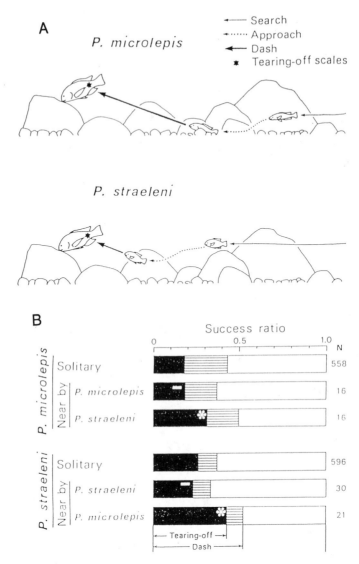

Figure 1. Hunting behaviour and success ratio of two scale eaters, *Perissodus microlepis* and *P. straeleni* in Lake Tanganyika. (A) Sequence of hunting behaviour of the species. (B) Hunting success ratio of the two species under the different foraging situations. * or - indicate that the rates are significantly different or not from that in a solitary feeding situation at the 5% level, respectively. (from Hori, 1987).

The other example is the 'cry substance' of plants. A leaf of a plant, when attacked by a species of herbivore, releases particular chemical substances, not in direct defence to the herbivore, but, instead, to call a carnivore to attack the herbivore (Dicke et al., 1990; Takabayashi, 1992; Takabayashi and Dicke, 1993). Specific chemicals are released only when a particular species is bitten: i.e., different substances are released to call different carnivores (Takabayashi, 1992, 1995).

Thus, these are two very 'simple' examples, from which it can be easily understood how complex and sophisticated the interactions among life forms are.

From Ecosystems to Genes, or "Symbiosphere: Ecological Complexity for Promoting Biodiversity"

In 1992, a book was published by the International Union of Biological Sciences (IUBS), the Scientific Committee on Problems of the Environment (SCOPE), and the United Nations Educational, Scientific and Cultural Organization (UNESCO), "From Genes to Ecosystems: A Research Agenda for Biodiversity" (Solbrig, 1992). I am inclined, however, to add to the book a viewpoint, 'from ecosystems to genes'.

Accumulated evidences strongly suggest that ecological complexity, the complex and sophisticated interactions among various forms of life, together with intricate heterogeneous habitat structure, plays a key role in promoting biodiversity in nature (Kawanabe et al., 1993).

A clear example of external inputs promoting biodiversity can be seen in the immunological responses of an organism against non-self materials invading the body. The diversity in antibodies based on the diversity of restructured DNA sequences produced in immunological cells is based on the necessity to counter the diversity in potential antigens present in the environment.

Plant and animal interactions offer another rich source of analogous examples. Plants develop defense systems against attack by a variety of herbivores. The hard structure made of cellulose, lignin and other cell-wall substances serves as a basic physical defense. Secondary substances which plants produce and store in their cytoplasm are effective chemical defenses. Animals attacking plants also develop a means for coping with these defenses. This coevolutionary process may lead to the creation of amazing products and the ability to produce them on both sides. This is an example of the case in which species interactions, through evolutionary processes, generate novel biological traits, thereby enhancing biodiversity.

Recent advances in community ecology extend this pairwise scheme of species interactions into that of more complex interactions, involving a third organism. 'Cry substances' released by plants, as mentioned above, are an excellent example of this phenomenon. Furthermore, the attacked plants may affect uninfested neighbouring plants, causing them to initiate defensive reactions (Bruin, et al., 1991). Even insects feeding at different times or on different parts of a plant may have substantial effects on the quality of resources available to one another (Ohgushi, 1992). Such indirect effects, mediated by the host plant, are more common than previously thought.

It is now clear that if such interactions had not occurred in evolutionary processes nor had not continued to the present, such genetic mutations would not be realized and would not have been maintained up to now. It also begins to show clearly that at the present time, the variability and flexibility of species interactions, or 'facultative interactions', and indirect effects among community components have also increased biodiversity at the gene and species levels.

Ecological complexity or complex and sophisticated ecological interrelationships, enhances biodiversity through evolutionary and biogeographical processes. In contrast, its degradation should quickly cause diversity in phenotypic and genotypic traits to decline.

For the conservation of biodiversity, it is not sufficient to preserve living organisms or their gametes alone, because keeping animals and plants in zoological and botanical gardens or their gametes in the frozen state cannot conserve the full range of diversity they exhibit in nature, due to the loss of the ecological complexity they enjoy in their original habitats. Moreover, to promote richer biodiversity in the future, we should encourage more complexity in biological communities and their present environments than exists at present.

Thus we proposed a project, 'Symbiosphere: Ecological Complexity for Promoting Biodiversity' in the 'DIVERSITAS', IUBS-SCOPE-UNESCO programme on biodiversity. It appeared in the "Biology International Special Issue," the proceedings of an international workshop held at Kyoto the previous year (Kawanabe et al., 1993).

The Importance of Functional Diversity of Microbial Organisms

In the case of higher plants and animals, the simplistic view of biodiversity is the species. First, because they are visible and can be easily separated from each other to individual species. In this case, however, the diversity of their functions, or more strictly speaking, the interrelationship among organisms and between organisms and their environments, have been considered at a secondary level. The 'one species-one niche' concept is one of the most symbolic examples. Larvae and imagos of a species of butterfly, for instance, have completely different functions in a community, i.e., they belong to completely different niches. Comparing this difference, imagos of congenetic but different species have very similar functions with each other (Kawanabe, 1982). Consideration of the functional units, or niches, is still at a very primitive stage for such macroorganisms.

In contrast, microorganisms are not readily visible and microbial species can't be easily determined, in many cases. This situation is, of course, one of the weakest points in microbial studies. It is a somewhat critical point for microorganisms than for macroorganisms, however, because functional units or niches of the former are much more easily observed.

In some lakes, there is a fraction of the microorganisms, which continuously keep a very small biomass, the function of which is still unclear. Such groups of microorganisms might have a very important role in the entire community or ecosystem.

The keystone species concept was first proposed by Paine (1969) and was a theory that was well known and admired, but was subsequently criticized by many ecologists. I believe, however, that key functions or key relationships play a very important role in the community, in general, and I expect it will be discovered to be so in microbial systems.

Macro-Microbial System as the Real Unit of Biodiversity

Symbiotic relationships between organisms represent a very common phenomenon, more than previously suspected, e.g. symbiotic feeding of Tanganyikan cichlids, as mentioned above. It is also clear at the present time that the 'competitive exclusion principle' is realized in many cases by the presence of two species, but is not, when a third species is present, as was discovered by field studies and laboratory experiments, as well as by mathematical analyses (Higashi, 1993, 1995; Higashi and Ohgushi, 1990).

The classical definition of the term 'symbiosis', especially as given by British ecologists, was restricted to lichens and corals or certain molluscs and the zooxanthellae. In a somewhat wider case, it includes 'digestive symbiosis', i.e., microorganisms in the termite gut, rumen, or mammalian blind gut.

Symbiotic relationships between macro- and microorganisms, however, are quite common and are very important in most communities. For example, there are very few or virtually none among the woods, or more broadly, the plants without symbiotic microorganisms.

Most scientists have recognized obligatory relationships, not only between macro- and microorganisms but also between macroorganisms. It is now clear, however, that facultative relationships among organisms are more common and probably much more important in communities or ecosystems than was realized. Interrelations among macro- and microorganisms should be investigated more intensively and extensively in the future.

I recall a part of a paper which was read at a meeting of the British Ecological Society, as a presidential address in 1948 by Charles S. Elton (1949). In the middle of the paper on

"Population interspersion: an essay on animal community patterns', he stressed relationships among animals, dead wood, and fungi, and then wrote, as follows: "Perhaps we may eventually have ecological survey lists that give all the fungi in detail by species, and mention at the bottom just 'Phanerogams and mosses also present.'"

The title of this symposium is 'Microbial diversity'. The study of microbial diversity, as such, mabye out of date or without meaning, even as the study of macrobial diversity, from an ecological point of view. It would be, and should be, diversity studies of macro-microorganism systems, in the case of the latter.

DIWPA: International Network of Diversitas Western Pacific and Asia

The DIWPA project is described very briefly as follows. Visualize the globe and imagine longitudinal belts from the north pole, across the equator, to the south pole. Throughout the world, you can find deserts, or grasslands at least, in the middle of every belts except one. It is situated in the Western Pacific and Asia, including the eastern end of Australia. Forests continue from Far East Russia, as boreal forests, to Japan, Korea and China, as temperate forests, then the Philippines, Indochina, Thailand, Malaysia, Indonesia and Papua New Guinea, as tropical forests, to the eastern end of Australia and New Zealand, as warm and cool temperate forests.

I would like to add another condition in this belt, the tallest tropical rainforests. No glacier cover was present, even in the northern part, e.g., the Lake Baikal area had not been completely covered by ice during the last glacial period. This situation also leads to higher biodiversity in terrestrial communities as a whole in this belt.

High biodiversity is observed in aquatic ecosystems, both freshwater and marine, along this belt, as well.

Thus, the 'International Network of *DIVERSITAS* in West Pacific and Asia', abbreviated as 'DIWPA', was proposed and established at the international symposium, 'Ecological Perspective of Biodiversity', dedicated, in part, to the 'International Prize of Biology', held in Kyoto in December, 1993.

At the IUBS Assembly, the 'International Forum of Biodiversity' held in Paris last September, the network was officially initiated. Workshops and symposia on biodiversity in Western Pacific and Asia at Singapore were held December, 1995, at Beijing in May, 1996, and will be held at Fiji in July, 1997, respectively. I hope microbial ecologists will have an interest in this network.

ACKNOWLEDGMENT

I thank Usio Shimizu for giving me the opportunity to read the paper in the symposium and thank Usio Simidu, Kenji Kato, Masahiko Higashi and Rita Colwell for correcting the manuscripts.

REFERENCES

Bruin, J., Sabelis, M. W., Takabayashi, J. and Dicke, M., 1991. Uninfested plants profit from their infested neighbours. *Proc. Exper. and Appl. Entomol., N. E. V. Amsterdam*, 2: 103-108.

Dicke, M., Sabelis, M. W., Takabayashi, J., Bruin, J. and Posthumus, M. A., 1990. Plant strategies of manipulating predator-prey interactions through allelo-chemicals: prospects for application in pest control. *J. Chem. Ecol.*, 16: 3091-3118.

Elton, C. S., 1949. Population interspersion: an essay on animal community patterns. *J. Ecol.*, 37: 1-23.

Higashi, M., 1993. An extension of niche theory for complex interactions. In: Kawanabe, H., Cohen, J. E. and Iwasaki, K. (ed) *Mutualism and community organization: behavioural, theoretical and food web approaches,* 311-322. Oxford Univ. Pr.

Higashi, M., 1995. Mechanisms for coexistence of various organisms. *Creative World*, 93: 30-49. (in Japanese)

Higashi, M. and Ohgushi, T., 1990. Three interconnections in ecology. In: Kawanabe, H., Ohgushi, T. and Higashi, M. (ed), Ecology for tomorrow. *Physiol. Ecol.*, 27 (Spec. No.): 199-205.

Hori, M., 1983. Feeding ecology of thirteen species of *Lamprologus* coexisting at a rocky shore of Lake Taganyika. *Physiol. Ecol.,* 20: 129-149.

Hori, M., 1987. Mutualims and commensalism in the fish community of Lake Tanganyika. In: Kawano, S., Connell, J. H. and Hidaka, T. (ed) *Evolution and coadaptation in biotic communities.* 219-239. Univ. Tokyo Pr.

Hori, M., 1991. Feeding relationships among cichlid fishes in Lake Tanganyikia: effects of intra- and interspecific variations of feeding behaviour on their coexistence. *Ecol. Intern. Bull.* 19: 89-101.

Hori, M., 1993. Frequency-dependent natural selection in the handedness of scale-eating cichlid fish. *Science*, 260: 216-219.

Kawanabe, H., 1982. Speciation would be occurred in particular life history stages. *Science Tokyo*, 52:527-529. (in Japanese)

Kawanabe, H., Ohgushi, T. and Higashi, M. (ed), 1993. Symbiosphere: ecological complexity for promoting biodiversity. *Biol. Intern., Spec. Issue*, 29. 86 p. Intern. Union Biol. Sci.

Kohda, M., 1994. Interspecific societies in cichlid fishes: inter- and intraspecific relations in herbivores. In: Hori, M. (ed) *Fishes in Lake Tanganyika: an attempt to solve a riddle of biodiversity,* 144-160. Heibonsha. (in Japanese)

Nakai, K., 1993. Foraging of brood predators restricted by territoriality of substrate-brooders in a cichlid fish assemblage. In: Kawanabe, H., Cohen, J. E. and Iwasaki, K. (ed) *Mutualism and community organization: behavioural, theoretical and food web approaches,* 84-108. Oxford Univ. Pr.

Ohgushi, T. (ed) 1992. *Various type of symbiosis: diversified interactions among species.* 230 pp. Heibonsha. (in Japanese)

Paine, R. T., 1969. A note on trophic complexity and community stability. *Amer. Natur.*, 103: 91-93.

Solbrig, O., 1992. *From genes to ecosystems: a research aggenda for biodiversity.* 4+124 pp. IUBS, SCOPE and UNESCO.

Takabayashi, J., 1992. Plants which employ body-guards: new picture of biological world brought by chemistry. In: Higashi, M. and Abe, T. (ed), *What is symbiotic earth ?,* 184-189. Heibonsha. (in Japanese)

Takabayashi, J., 1995. Reproductive strategy of plants and symbiosis. *Creative World,* 93: 12-29. (in Japanese)

Takabayashi, J. and Dicke, M. 1993. Volatile allelochemicals that mediate interactions in a tritrophic system consisting of predatory mites, spider mites, and plants. In: Kawanabe, H., Cohen, J. E. and Iwasaki, K. (ed) *Mutualism and community organization: behavioural, theoretical and food web approaches,* 280-295. Oxford Univ. Pr.

Takamura, K., 1983. Interspecific relationship between two aufwuchs eaters *Petrochromis polyodon* and *Tropheus moorei* of Lake Tanganyika, with a discussion on the evolution and functions of a symbiotic relationship. *Physiol. Ecol.,* 20: 59-69.

Yuma, M., 1993. Competitive and co-operative interactions in Lake Tanganyika fish communities. In: Kawanabe, H., Cohen, J. E. and Iwasaki, K. (ed) *Mutualism and community organization: behavioural, theoretical and food web approaches,* 213-227. Oxford Univ. Pr.

Yuma, M., 1994. Food habits and foraging behaviour of benthivorous cichlid fishes in Lake Tanganyika. *Env. Biol. Fishes*, 39: 173-182.

3

FROM STRAINS TO DOMAINS

Measuring the Degree of Phylogenetic Diversity among Prokaryotic Species and as Yet Uncultured Strains

Erko Stackebrandt

DSM-German Collection of Microorganisms
    and Cell Cultures GmbH
Mascheroder Weg 1b
38124 Braunschweig, Germany

INTRODUCTION

Two molecular methods have proven successful in determining the whole range of relationships between prokaryotic organisms, i.e. from the level of strains to the level of domains. DNA:DNA hybridization is suitable for the elucidation of relationships among closely related strains, while all other levels, from remote to distant relationships, can be unravelled by the analysis of conserved macromolecules, preferably the RNA of the small subunit of ribosomes or the genes coding for it (16S rRNA/DNA) (Stackebrandt and Goebel, 1994). It is obvious that DNA reassociation methods require pure cultures, in order to determine the degree of relatedness between individual organisms, while 16S rDNA sequences can be obtained from mixed bacterial populations by analysis of cloned DNA restriction fragments or PCR products (Giovannoni et al., 1990; Ward et al., 1990). The presence of rDNA in every organism, and the availability of a broad range of molecular methods, including amplification techniques and detection methods, e.g., the use of taxon- specific PCR primers and oligonucleotide probes, makes sequence analysis and/or detection of 16S rDNA the superior approach for the assessment of phylogenetic diversity (Pace et al., 1985). The analysis of this molecule does not allow determination of morphological and physiological activities, hence information about the ecological role played by an organism in its environment; but it reveals information about approximate relationships between naturally occurring organisms and cultured strains, and it allows us to determine the presence of novel types of prokaryotes.

Microbial Diversity in Time and Space, edited by Colwell et al.
Plenum Press, New York, 1996

THE DEMANDING TASK OF DESCRIBING A PROKARYOTIC SPECIES

One of the most interesting problems that cannot be answered directly is whether clone sequences can be assigned to taxa presently defined in bacteriology. In this context, it is important to recapitulate the species definition in bacteriology. The following problems have been recognized that hamper in bacteriology the development of a species concept, comparable to those applied to the majority of eukaryotic species.

1. Convincing data are lacking that allow us assume that evolution leads to genetically and phenotypically isolated prokaryotic entities. Considerable overlap exists between closely related taxa at the level of the primary structure of DNA, as determined by DNA reassociation studies and derived from the high degree of shared phenotypic characters.

2. Due to the relatively high mutation rate, a bacterial colony represents a mixture of genetically and phenotypically different strains that cannot, in most cases, be detected without extensive genotypic characterization.

3. The genotype of a bacterial culture may be altered through the processes of conservation and possibly by other physico-chemical parameters, which ultimately may alter the phenotype as well.

4. The ecological niche is often not known and closely related strains have been isolated from quite different habitats.

5. It is likely that some properties that are expressed in the natural environment are lost when the isolates are maintained under quasi-artificial conditions in the laboratory.

6. The extent to which gene exchange occurs between closely related strains is unknown for the majority of cultured strains. Gene exchange through horizontal and vertical gene transfer may change quite dramatically the phenotypic properties of genetically closely related strains, including those characters which are defined to be taxon-specific.

Because it is presently impossible to determine whether there exists in nature a prokaryotic entity that is comparable to that defined in zoology and botany, the prokaryotic species definition combines features of both the genotype and the phenotype, hence following a polyphasic approach to taxonomy (Colwell, 1970). As taxonomic history has shown that phenotypic properties alone are inadequate for the definition of a species, this taxon is currently defined as a group of strains which are highly related, with respect to the primary structure of their DNA (as determined by about 70% reassociation values), and share a high degree of phenotypic similarity (Wayne et al., 1987). Gene amplification, as well as gain and loss of genes, have little influence on species delineation, as long as the DNA reassociation threshhold value is not significantly lowered by these processes, and the phenotype is not altered in those characters used in the species description. Thus, the species definition is rather robust, allowing for a number of changes at the genomic and phenotypic level. However, the decription of a new species is rather demanding, as determination of chemotaxonomic properties and DNA reassociation studies are time-consuming and laborious. This explains why the two prokaryotic domains, Archaea and Bacteria, presently contain only 3200 validly described species (excluding cyanobacteria), belonging to 204 genera (October, 1994). Over the last 15 years, the yearly increase in number of new species has ranged between 70 and 150, and that of new genera between 15 and 30. Compared to the total number of described biological species, the fraction of prokaryotic species is only about 0.2 percent. This appears unlikely, as prokaryotes evolved about 3.8 billion years ago and are, therefore, significantly older than those species known to represent species-rich taxa (e.g., insects [~69%] and plants [~14%]) (Hammond, 1992), but are also able to thrive and evolve in niches that cannot be occupied by eukaryotes.

MOLECULAR ENVIRONMENTAL STUDIES HAVE REVEALED NEW INSIGHTS INTO THE PHYLOGENY OF HITHERTO UNCULTURED ORGANISMS

One prerequisite for the assessment of microbial diversity in natural environments is the availability of an extensive molecular database of cultured organisms which can serve as a reference for comparison of sequences from both isolates and uncultured strains. Comparison of sequences from environmental rDNA clone sequences to each other, and to cultured strains allows the recognition of putative target sites for oligonucleotides, suitable for specific detection of the respective strains directly in their natural habitat. The 16S rDNA database is already so extensive that about 90% of all described species can be placed rather accurately within the radiation of the main lines of descent.

Although, at present, expectations are not very high, it is hoped that eventually the conditions required for cultivating at least those strains that are highly related to cultured strains will be determined. Furthermore, the assemblage of sequences from different natural sites for which physico-chemical parameters have been determined, will eventually add to the understanding about the role these organisms play in their habitats and about the ecological forces that have selected for these strains.

Studies of 16S rDNA sequences obtained from clone libraries generated from PCR-amplified DNA, retrieved from different habitats, and from different parts of the world, can be summarized as follows (due to space limitations only a few references are included):

1. Methodological problems exclude the possibility of quantifying taxa determined to be present in clone libraries; these problems are associated with the extraction of nucleic acids (Liesack and Stackebrandt, 1992), PCR-primer sensitivity and selectivity (Stackebrandt et al., 1993), cloning steps (Rainey et al., 1994), and the dependence of PCR amplificate amount on undeterminable genomic properties (Farelly et al., 1995). Thus, it is also not possible to decide whether the identified clones belong to a majority or a minority population of the naturally occurring prokaryotes.

2. Some sequences obtained from samples collected at widely separated geographical locations show surprisingly high similarities. This finding is based mainly on results obtained for samples collected from the Pacific and Atlantic Oceans (Fuhrman et al., 1993; Giovannoni et al., 1990), but some examples have also been found in samples from soil habitats (Liesack and Stackebrandt, 1992; Ogram and Bollinger, unpublished).

3. The vast majority of retrieved sequences are not identical to those of cultured strains; exceptions include some marine cyanobacteria and acidophilic strains from a commercial bioleach reactor (Goebel and Stackebrandt, 1994).

4. While most bacterial sequences can be assigned to known phyla, archaeal sequences have been retrieved from marine environments (including Antarctica [DeLong et al., 1994]) that point towards completely novel kingdoms (Barns et al., 1994; DeLong, 1992; Fuhrman et al., 1992)

5. Sequences of strains isolated from the same geographical locations from which the DNA was retrieved for the generation of a clone library are rarely identical to

the clone sequences and also to those sequences available in the 16S rDNA database. An exception is the bioleach reactor study.

6. How should the occurrence of highly similar sequences that differ in a few nucleotides only (Britschgi and Giovannoni, 1991) be interpreted? Is this a reflection of rRNA operon microheterogeneity, or of the presence of highly related strains?

ASSESSMENT OF GENETIC DIVERSITY

Comparative 16S rDNA sequence analysis of pure cultures has shown that it is not possible to delineate ranks at any level purely on the basis of similarity or dissimilarity values. For example, the phylogenetic depth of the taxon "species" (higher than about 70% DNA reassociation) may range from 97% 16S rDNA similarity to absolute sequence identity. The degree of sequence similarity varies even more for higher taxa, i.e. between the levels of the genus and the phylum (kingdom). Consequently, it is not possible to deduce taxon affiliation of an isolate or a clone sequence on the basis of sequence similarity alone (Stackebrandt, 1991). The only approximate reference similarity value available at present is the 97% similarity value; as in no case did the corresponding genomic DNAs hybridize at greater than about 60%, it can be excluded that, below the 97% level, cultured or uncultured organisms belong to the same species.

To evaluate similarity values for allocation of clone sequences to species, it is important to stress that the 97% threshhold value was determined on the basis of complete 16S rDNA sequences. Restriction of the analysis to more conserved stretches or to more variable regions will lead to higher or lower similarity values, respectively. As the position of hypervariable regions differ in different groups of organisms it is difficult to estimate precisely the degree of similarity of complete sequences from partial sequences. Using selected sequence stretches of members of the genus *Vibrio*, as an example, i.e. region 50 through 500 and region 550 through 1500, the percent similarity is about 3% higher and 2% lower, respectively, when compared to the values obtained with complete sequences. In this genus the hypervariable regions are located in the 3′ half of the molecule, while in streptomycetes, for example, these regions are found at the 5′ half of the 16S rDNA molecule (Stackebrandt et al., 1992). Most environmental sequences, however, are shorter than about 1500 nucleotides, ranging in size between about 200 and 1200 nucleotides obtained from different positions of the molecule (Fig. 1). In order to allow comprehensive analysis of all available clone sequences of environmental DNA, it is recommended that researchers agree to determine routinely the sequence of at least the 5′ terminal 500 nucleotides. In many of those examples, where the degree of sequence similarity has been determined between clone sequences of length 1000 nucleotide and longer, generated from environmental DNA and their nearest cultured neighbor, the values are far below 95% similarity, often as low as only 85 to 80%.

Certainly, the majority of these values indicate that the uncultured strains represent novel genera, families and sometimes even classes. In almost no case were clone sequences, of whatever length, absolutely identical to those of culture collection strains, ranging between 99 and 97% similarity. In order to determine whether the, as yet, unknown strains are members of described species, the environmental strains must be isolated, and their phenotype, and if necessary their DNA similarity, be determined. The low numbers of absolutely matching sequences indicate that the vast majority of clone sequences indeed represent novel species in the environment. To obtain a more complete picture, it will be necessary to have a complete set of sequences from the type strains of all validly described species.

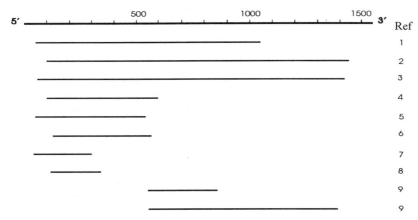

Figure 1. Regions of 16S rDNA used for obtaining information about phylogenetic diversity in environmental samples. References: 1, Stackebrandt et al., 1992, FASEB J. *7*:232-236; 2, De Long et al., 1993, Limnol. Oceanogr. *38*:924-934; 3, Britschgi and Giovannoni, 1991, Appl. Environ. Microbiol. *57*:1707-1713; 4, Choi et al., 1994, Inf. Immun. *62*:1889-1895; 5, Schmidt et al., 1991, J. Bacteriol. *173*:4371-4378; 6, Ward et al., 1990, Nature *345*:63-65; 7, Liesack et al., 1992, Biodiv. Conserv. *1*:250-262; 8, Giovannoni et al., 1990, Nature *345*:60-63; 9, Fuhrman et al., 1993, Appl. Environ. Microbiol. *59*:1294-1302.

The molecular approach, is only one, but an important step in the elucidation of ecological activities in natural samples. Such studies must be accompanied by (i) the identification of cells in the environment through *in situ* probing, (ii) determination of cell activities through application of "functional" probes, (iii) determination of the metabolically active part of the population by analysis of mRNA, (iv), and attempts to isolate the, as yet, uncultured strains.

Table 1. Examples of phylogenetic distance of 16S rDNA clones to their nearest culturable neighbor (in percent similarity)

Clone designation	Nearest phylogenetic neighbor	Distance to neighbor	Reference in Fig. 1
NZM 3143	*Treponema pallidum*	~30%	4
NZM 3101	*Treponema succinifaciens*	~17%	4
MC31	*Verrucomicrobium spinosum*	~25%	1
MC58	Iron oxidizer TH3	~10%	1
MC103	New actinomycete line	~20%	1
MC55	*Gemmata obscuriglobus*	~18%	1
MC100	*Planctomyces limnophilus*	~14%	1
ALO39/SAR11	*Caulobacter crescentus*	~16%	5/8
ALO7/SAR6	*Synechococcus species*	~6%	5/8
ALO40	*Chromatium vinosum*	~7%	5
MC6	*Bradyrhizobium japonicum*	~2%	7
MC2	*Rhodopseudomonas palustris*	~1%	7

REFERENCES

Barns, S.M., Fundyga, R.E., Jeffries, M.W., and Pace N.R., 1994, Remarkable archaeal diversity detected in a Yellowstone National Park hot spring environment. Proc. Natl. Acad. Sci. USA *91*:1609-1613.

Britschgi, T., and Giovannoni, S. J., 1991, Phylogenetic analysis of a natural marine bacterioplankton population by rRNA gene cloning and sequencing. Appl. Environ. Microbiol. *57*:1707-1713.

Colwell, R. R., 1970, Polyphasic taxonomy of bacteria. *In* Proceedings of the International Conference on Culture Collections, ed. H. Iizuka, T. Hasegawa, pp. 421-436. Tokyo:Tokyo Unversity Press.

De Long, E.F., Wu, K.Y., Prezelin, B.B., and Jovine, R.V.M., 1994, High abundance of Archaea in Antarctic marine picoplancton. Nature *371*:695-697.

DeLong, E. F., 1992, Archaea in coastal marine environments. Proc. Natl. Acad. Sci. USA. *89*:5685-5689.

Farelly, V., Rainey, F.A., and Stackebrandt, E., 1994, Effect of genome size and *rrn* gene copy number on PCR amplification of 16S rRNA genes from a mixture of bacterial species. Appl. Environ. Microbiol. *61*: 2798-2801.

Fuhrman, J. A., McCallum, K., and Davis, A. A., 1992, Novel major archaebacterial group from marine plankton. Nature *356*:148-149.

Fuhrman, J. A., McCallum, K., and Davis, A. A., 1993, Phylogenetic diversity of subsurface marine microbial communities from the Atlantic and Pacific oceans. Appl. Environ. Microbiol. *59*:1294- 1302.

Giovannoni, S. J., Britschgi, T. B., Moyer, C. L., and Field, K. G., 1990, Genetic diversity in Sargasso Sea bacterioplankton. Nature *345*: 60-63.

Goebel, B.M., and Stackebrandt, E., 1994, Cultural and phylogenetic analysis of mixed microbial populations found in natural and commercial bioleach environments. Appl. Environ. Microbiol. *60*:1614-1621.

Hammond, P.M., 1992, Species inventory. *In* Global biodiversity: Status of the earth's living resources, pp. 17-39, ed. B. Groombridge. London: Chapman and Hall.

Liesack, W., and Stackebrandt, E., 1992, Occurrence of novel groups of the domain *Bacteria* as revealed by analysis of genetic material isolated from an Australian terrestrial environment. J. Bacteriol. *174*:5072-5078.

Pace, N.R., Stahl, D.A., Lane, D.J., and Olsen, G.J., 1985, The analysis of natural microbial communities by ribosomal RNA sequences. Microb. Ecol. *9*:1-56.

Rainey, F.A., Ward, N., Sly, L.I., and Stackebrandt, E., 1994, Dependence on the taxon composition of clone libraries for PCR amplified, naturally occurring 16S rDNA, on the primer pair and the cloning system used. Experientia *50*:796-797.

Stackebrandt, E, 1991, Unifying phylogeny and phenotypic diversity. *In* The Prokaryotes, ed. A. Balows, H.G. Trüper, M. Dworkin, W. Harder, and K.-H. Schleifer, pp. 19-47. New York: Springer.

Stackebrandt, E., and Goebel, B.M., 1994, Taxonomic Note: A place for DNA-DNA reassociation and 16S rRNA sequence analysis in the present species definition in Bacteriology. Int. J. Syst. Bacteriol. *44*:846-849.

Stackebrandt, E., Liesack, W., and Witt, D., 1992, Ribosomal RNA and ribosomal DNA sequence analysis. Gene *115*:255-260.

Stackebrandt, E., Liesack, W., and Goebel, B.M., 1993, Bacterial diversity in a soil sample from a subtropical Australian environment as determined by 16S rDNA analysis. FASEB J. *7*:232-236.

Ward, D.M., Weller, R., and Bateson, M.M., 1990, 16S rRNA sequences reveal numerous uncultured inhabitants in a natural community. Nature *345*:63-65.

Wayne, L., Brenner, D.J., Colwell, R.R., Grimont, P.A.D., Kandler, O., Krichevsky, M.I., Moore, L.H., Moore, W.E.C., Murray, R.G.E., Stackebrandt, E., Starr, M.P., and Trüper, H.G., 1987, International Committee on Systematic Bacteriology: Report of the ad hoc committee on reconciliation of approaches to bacterial systematics. Int. J. Syst. Bacteriol. *37*:463-464.

4

DIVERSITY OF VIRUSES

Masakazu Hatanaka

Institute for Virus Research
Kyoto University
Kyoto 606-01, Japan

INTRODUCTION

Viruses, like other living organisms, receive their genetic information from the past, maintain and transmit the information in the present, and replicate and send the information faithfully to time and space of the future.

However, unlike other organisms, viruses achieve their function only after entering into a living organism, a host. Therefore, viral diversification depends upon selection of host, mode of transmission, and strategies of replication, transcription and translation within a cell.

SELECTION OF HOST

Viruses select their hosts for replication, e.g., the Influenza A virus can replicate in the host of water birds, pigs or human beings. The virus contains eight segments of RNA in a viral particle that may mutate easily (drift) and exchange segments (shift) so that periodically the virus continues to proliferate, despite of the presence of host immunity arising against the same virus.

To replicate, a virus must enter into the host cell through its virus receptor, which is scattered over the cellular membrane surface.

A virus freely chooses its own receptor, which includes transporters of amino acids or ions, lipids, and CD4 antigen, among others. There is no rationale for selecting a membrane molecule, since any viral receptor may promote, in a way, diversification of viruses.

Some viral genomes are integrated into the host genome (McAllister et al., 1972; Okabe et al., 1973a,b). The retrovirus, transmitted through gametes, is called an endogenous retrovirus. Endogenous retroviruses have been found in the genomes of mice, rats, cats, and baboons (Scolnick et al., 1974; Tsuchida et al., 1974a, 1975). We can induce infectious viruses from virus-free cells by adding analogues of nucleosides or amino acids to the culture media (Gilden et al., 1974; Tsuchida et al., 1974b).

Microbial Diversity in Time and Space, edited by Colwell et al.
Plenum Press, New York, 1996

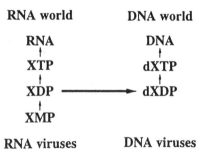

Figure 1. Molecular evolution of viruses.

There is diversity among retroviruses, for example, an endogenous retrovirus derived from mouse utilizes only the mouse as host organism and is referred to as an ecotropic virus. Some cat endogenous retroviruses proliferate only in species other than the cat (McAllister, et al., 1972) and are called xenotropic viruses. The retroviruses, which replicate in two different species and exist as a mouse endogenous virus are amphotropic. The viruses, therefore, are transmitted horizontally and/or vertically through intra-species, inter-species and/or trans-spieces hosts (Kurstak and Maramorosch, 1974).

MOLECULAR EVOLUTION OF VIRUSES

Before DNA organisms arose about 4 billion years ago, the first genetic information may have been coded in RNA, and then transferred to DNA. Presently we see replication of RNA only in the RNA viruses (Fig. 1). Also reverse transcription of genetic information, from RNA to DNA is found only in retroviruses (Fig. 2).

The viruses produce massive quantities of progeny at a time by multiple replication of viral genes during the eclipse phase in a host cell. This causes variation in the viral genome, particularly when a virus has no correcting or repair enzyme. Most RNA viruses have no capacity for repair and, as a result, errors in genetic replication are significant. HIV reverse transcriptase, for example, synthesizes DNA with one base error our ot 600 bases, when viral RNA is used as a template. This causes many variants among the viral progeny. RNA viruses mostly contain single stranded RNA, the replication of which is different from that of double

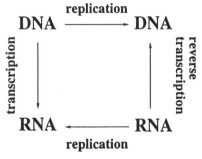

Figure 2. RNA replication and reverse transcription of genetic information by viruses.

RNA virus

dsRNA(±)   sense and antisense
ssRNA(+)   sense
ssRNA(−)   antisense
ssRNA(+ and −)   ambisense
ssRNA(0)   nonsense

Figure 3. Genetic information of RNA viruses. ds: double strand. ss: single strand.

stranded DNA found in all living organisms. DNA organisms replicate with semi-conservative double stranded DNA, while virus with single stranded DNA or RNA cannot conserve faithfully the genetic information. This causes genesis of polymorphism or genotypic variants (Siomi et al., 1988a).

Single stranded DNA or RNA viruses contain sense or antisense genes. In addition, some RNA viruses contain an ambisense strand which consists of partly sense and partly antisense within one RNA strand in a mosaic. In an extreme case, viroids retain apparently no coding information (Fig. 3).

DIVERSITY OF REPLICATION AND TRANSCRIPTION

All living organisms use RNA as a primer for DNA replication, again suggesting creation of the RNA world first, followed by the DNA world. On the other hand, viruses use a variety of strategies for replication of genetic information.

Although most viruses use RNA as a primer for DNA synthesis (Okabe et al., 1972), some utilize DNA or protein as a primer, and others need no primer for replication of the viral genome. Viruses have their own strategies for genetic replication that drive biodiversity of viral species.

Some viruses facilitate their replication through activation of host transcription, with retrovirus replication intimately coupled with the transcription machinery of their host. The reverse-transcribed virus information is integrated into the host genome and their propagation and progeny formation depend upon the transcription complex of the host cells. Some complex retroviruses have evolved by acquiring trans-activating gene(s) and their cis element(s) (Fig. 4) (Haseltine and Wong–Staal, 1991).

After transcription, the viral RNA is modified to form several molecular species of RNA, consisting of unspliced, partially spliced, and fully spliced RNA molecules. Although eukaryotic cells cannot transport unspliced nuclear RNA to the cytoplasm, unspliced RNA of viral origin is transported from nucleus to cytoplasm, where one is used as messenger RNA of gag, pro and pol genes, another is moved near the cellular membrane to become a viral gene within a viral particle (Siomi et al., 1988b; Nosaka et al., 1989). Partially spliced RNA becomes env mRNA, and fully spliced RNA forms mRNAs of regulatory genes.

Trans-activation

Cis-element

Host factors

Figure 4. Molecular diversity of virus transcription.

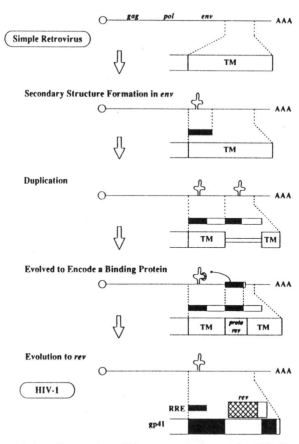

Figure 5. Molecular evolution of a retrovirus. TM: transmembrane protein. RRE: Rev responsive element. rev: RNA binding protein.

Partitioning of these viral RNAs is accomplished by the action of a regulatory protein of complex retrovirus, rex of HTLV-1 and rev of HIV-1, for example (Haseltine and Wong–Staal, 1991).

The complex virus with the genes of transcriptional and post-transcriptional regulation may have evolved at the molecular level from simple retrovirus by gene duplication (Fig. 5, Kubota et al., 1994).

TRANSLATIONAL DIVERSITY

Protein synthesis from messenger RNA starts from the first initiation codon near the 5' end of messenger RNA with the 5' cap structure. Influenza virus cuts the 5' ends of host pre-messenger RNAs and conjugates the cap-containing 10 nucleotides to the viral coding RNA to form the viral messenger RNAs.

Poliovirus takes a different strategy for viral protein synthesis. The virus uses no cap as a viral messenger RNA, instead, forms a unique structure called IRES or internal ribosomal entry site for protein synthesis (Fig. 6).

Cap

IRES
(Internal Ribosomal Entry Site)

Ribosomal Frameshift

Figure 6. Translational diversity of viruses.

The messenger RNA of eukaryotic cells usually contains only one coding frame and one translated product. The virus, on the other hand, exploits its limited nucleotides to translate multiproteins from one messenger RNA.

Most retroviruses perform ribosomal frameshifts to produce reverse transcriptase that is essential for their replication (Hatanaka and Nam, 1989a,b,c). In the case of HTLV-1, a double frameshift is required to produce the reverse transcriptase (Nam et al., 1993). See Fig. 7. From one mRNA, three protein molecules are produced, namely, gag, gag-pro, and gag-pro-pol gene products, by shifting coding frames minus one, each time (Hatanaka and Nam, 1989c).

BIODIVERSIFICATION BY VIRUSES

The occurrence of gene duplication or redundancy is suggested to be a driving force for diversification and evolution. Evidence that a retrovirus plays a role in duplicating nucleotide sequences in a host is shown in Fig. 8 (Kubota et al., 1993). This type of duplication mechanism arises in any gene or sequence in the host of a virus, which may have, very likely, contributed to increased biodiversity.

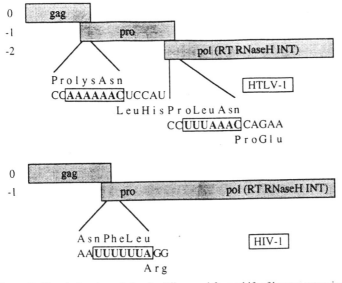

Figure 7. Translational regulation by Ribosomal frameshift of human retroviruses.

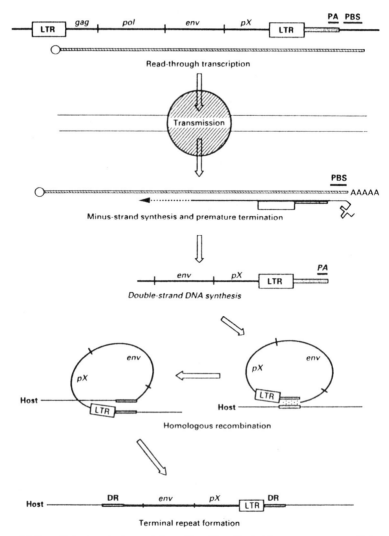

Figure 8. A model for cellular terminal repeat formation. The stippled box represents a cellular repeated sequence found in the tumor cell. Hatched line, with a circle at the head, indicates a read-through genomic RNA. Abbreviations displayed are as follows: PA, cellular polyadenylation signal; PBS, cellular primer binding site; DR, direct repeat. Small structured illustration below the PBS denotes a tRNA or tRNA-like nucleic acid as a primer. Arrows with bold lines (→) represent DNA synthesis. Open box indicates the transcripts over LTR.

CONCLUSION

Viruses evolved through interaction with their host organisms, whereas their hosts evolved restriction enzymes and immune systems through interaction with viruses. While some viruses integrated into the genome of their hosts, to become "self" of the host, some viruses and their hosts cohabitated together as non-self, without eliminating each other. All of the strategies of hosts and viruses aimed at transfer between host and virus may have contributed to the genesis of biodiversity, in which genetic information in the biomass evolved at the molecular level in all living organisms, together with viruses, for 4 billion years.

REFERENCES

Gilden, R.V., Oroszlan, S., and Hatanaka, M., 1974, Comparison and evolution of RNA tumor virus components, In Viruses, Evolution, and Cancer, ed. E. Kurstak, K. Maramorosch, pp.235-257. New York: Academic Press.

Haseltine, W.A., and Wong–Staal, F., 1991, Genetic Structure and Regulation of HIV. New York: Raven Press Ltd.

Hatanaka, M, and Nam, S-H., 1989a, Viral proteinases as targets for chemotherapy. Processing sequence of gag-pro polyprotein by HTLV-I proteinase, In Current Communications in Molecular Biology, ed. H.-G. Kräusslich, S. Oroszlan, E. Weimmer, pp. 117-126. New York: Cold Spring Harbor Laboratory press.

Hatanaka, M. and Nam, S-H., 1989b, Identification of HTLV-I gag protease and its sequential processing of the gag gene product. J. Cell. Biochem. 40: 15-30.

Hatanaka, M. and Nam, S-H., 1989c, Synthesis and activity of HTLV-I protease and its muteins. In Proteases of Retroviruses, ed. V. Kostka, pp. 119-124. Berlin: Walter de Gruyter.

Hatanaka, M. 1991, Nucleolar targeting signals (NOS) of HIV. In Genetic Structure and Regulation of HIV, ed. W.A. Haseltine, F. Wong-Staal, pp. 263-287. New York: Raven Press Ltd.

Kubota, S., Furuta, R., Maki, M., Siomi, H., and Hatanaka, M., 1993, Long cellular repeats flanking a defective HIV-I provirus: implication for site-targeted integration. Oncogene 8: 2873-2877.

Kubota, S., Oroszlan, S., and Hatanaka, M., 1994, The origin of human immunodeficiency virus type-1 rev gene: An evolutionary hypothesis. FEBS Lett. 338: 118-121.

McAllister, R.M., Nicolson, M., Gardner, M.B., Rongey, R.W., Rasheed. S., Sarma, P.S., Huebner, R.J., Hatanaka, M., Oroszlan, S., Gilden, R.V., Kabigting, A., and Vernon, L., 1972, C type virus released from cultured human rhabdomyosarcoma cells. Nature (New Biology) 235 : 3-6.

Nam, S.H., Copeland, T.D., Hatanaka, M. and Oroszlan, S., 1993, Characterization of ribosomal frameshifting for expression of pol gene products of human T-cell leukemia virus type I. J. Virol. 67: 196-203.

Nosaka, T., Siomi, H., Adachi, Y., Ishibashi, M., Kubota, S., Maki, M., and Hatanaka, M., 1989, Nucleolar targeting signal of human T-cell leukemia virus type I rex-encoded protein is essential for cytoplasmic accumulation of unspliced viral mRNA. Proc. Natl. Acad. Sci. USA 86: 9798- 9802.

Okabe, H., Lovinger, G.G, Gilden, R.V., and Hatanaka, M., 1972, The nucleotides at the RNA-DNA joint formed by the DNA polymerase of Rauscher leukemia virus. Virology 50: 935-938.

Okabe, H., Gilden, R.V., Hatanaka, M., 1973a, Extensive homology of RD-114 virus DNA with RNA of feline cell origin. Nature (New Biology) 244: 54-56.

Okabe, H., Gilden, R.V., and Hatanaka, M., 1973b, RD-114 virus- specific sequences in feline celluar RNA: Detection and characterization. J. Virology 12: 984-994.

Scolnick, E.M., Parks, W., Kawakami, T., Kohne, D., Okabe, H., Gilden, R., and Hatanaka, M., 1974, Primate Type C viral nucleic acid association kinetics: Analysis of model systems and natural tissues. J. Virology 13: 363-369.

Siomi, H., Nosaka, T., Saida, T., Miwa, H., Hinuma, Y., Shirakawa, S., Miyamoto, N., Kondo, T., Araki, K., Ichimura, M., Miura, A., and Hatanaka, M., 1988a, Two major subgroups of HTLV-I in Japan. Virus Genes 1: 377-383.

Siomi, H., Shida, H., Nam, S-H., Nosaka, T., Maki, M. and Hatanaka, M., 1988b, Sequence requirements for nucleolar localization of human T cell leukemia virus type I pX protein, which regulates viral RNA processing. Cell 55: 197-209.

Tsuchida, N., Gilden, R.V., and Hatanaka, M., 1974a, Sarcoma- virus-related RNA sequences in normal cells. Proc. Natl. Acad. Sci. USA 71: 4503-4507.

Tsuchida, N., Shih, M.S., Gilden, R.V., and Hatanaka, M., 1974b, Sarcoma and helper-specific RNA tumor virus subunits in transformed nonproducer mouse cells activated to produce virus by treatment with Bromodeoxyuridine. J. Virol. 14: 1262-1267.

Tsuchida, N., Gilden, R.V., Hatanaka, M., Freeman. A.E., and Huebner, R.J., 1975, Type C virus specific nucleic acid sequences in cultured rat cells. Int. J. Cancer. 15: 109-115.

5

THE EVOLUTION OF FUNGAL DIVERSITY

Past, Present, and Future

Amy Y. Rossman

Systematic Botany and Mycology Laboratory
USDA-Agricultural Research Service
10300 Baltimore Avenue
Beltsville, Maryland 20705

INTRODUCTION

Fungi are an heterogeneous assemblage of eukaryotic organisms united by an absorptive mode of nutrition in which they grow through the substrate, secreting degradative enzymes and absorbing nutrients. Because of this absorptive mode of nutrition, the fungi are ubiquitous, intimately associated with their substrates, and able to degrade a diverse range of organic and inorganic materials. The breakdown of substrates, primarily organic materials, by fungi results in the cycling of nutrients that is crucial for long term ecosystem maintenance. The relationships of fungi with their nutrient sources ranges from that of: *obligate parasites*, often causing serious diseases of plants, animals, and humans; *obligate mutualists*, as in the arbuscular mycorrhizal associations of the Glomales with roots of crops plants and in the anaerobic chytrids in the rumen of ruminant animals; *obligate commensals*, as in the Trichomycetes, a specialized group of fungi that exist in the hind guts of insects without causing any apparent harm to the host; to *facultative relationships*, primarily as *saprobes* degrading all kinds of plant material from simple sugars to the complex carbohydrates, lignin and chitin, often in specialized habitats, for example, yeasts fermenting flower nectar, as well as serving as *hosts* for viruses. Fungi often form complex communities interacting with other organisms inhabiting the same substrates. Thus, fungal diversity is reflected in the diversity of substrates available for fungal exploitation throughout evolutionary history and into the present and future.

PAST: EVOLUTIONARY ORIGIN OF THE FUNGI

Fungi, defined as those organisms studied by mycologists, are recognized as a diverse and unique group of organisms particularly following their separation from plants as the fifth kingdom (Whittaker, 1969). Their diversity continues to be elucidated, as an emerging fossil record is combined with biochemical, ultrastructural, and molecular evidence to answer

Microbial Diversity in Time and Space, edited by Colwell et al.
Plenum Press, New York, 1996

questions of evolution. Several conclusions about the origins of the fungi are well accepted. One is that the Oomycetes, having flagellated zoospores and exemplified by the history-making plant pathogen, *Phytophthora infestans*, causing late blight of potatoes, and other Peronosporales or downy mildews, should be included in the Chromista, a group previously considered to consist primarily of the Xanthophyceae or yellow-brown algae (Cavalier-Smith, 1987; Dick, 1988). Second, the Myxomycetes or slime molds and Labyrinthulales are considered to be allied with the Protoctista or protozoans (Cavalier-Smith, 1987). The other major groups of fungi, herein referred to as the true fungi, are the Chytridiomycetes, Zygomycetes, Ascomycetes, and Basidiomycetes.

The true fungi appear to share a common origin, with evidence suggesting that all major groups existed by the late Precambrian (Berbee & Taylor, 1993; Bruns, et al., 1992), yet many questions remain concerning both their origin and the relationships within this large and still rather diverse group of organisms. A most interesting conclusion, recently supported with several kinds of molecular data (Wainwright, et al., 1993; Baldauf & Palmer, 1993), suggests that true fungi are more closely related to animals than plants.

The ancient origin of the true fungi, suggested by molecular evidence, is supported by the fossil record. Although the literature is sparse, with many early reports now discredited, fossils for several groups of fungi bear a striking morphological similarity to extant fungal structures (Stubblefield and Taylor, 1988). Among the earliest known fungi are those associated with root-like structures of the earliest land plants in the Rhynie Chert of the Devonian (Pirozynski and Dalpé, 1989; Simon, et al., 1993). These fossil structures appear to be similar to vesicles occurring in modern arbuscular mycorrhizae. Recent papers describing Rhynie Chert fossils report and illustrate chytridiaceous fungal parasites on the vesicular structures that appear identical to extant parasites of mycorrhizal vesicles (Hass, et al., 1994; Taylor, et al., 1992). Sherwood–Pike (1991) and Sherwood–Pike and Gray (1985) present a case for the existence of fossil Ascomycetes in the Silurian, illustrating spores with the characteristic structure of some present day ascomycetous anamorphs. Fossil Basidiomycetes are known from both a definitive clamp connection, reported in the Pennsylvanian (Dennis, 1970), as well as fossilized wood that appears to be degraded in a manner morphologically characteristic of wood degraded by extant Basidiomycetes. Thus, both molecular evidence and the fossil record suggest a Precambrian origin of the true fungi with most major groups known from the late Precambrian.

PRESENT: MEASURING MODERN BIOLOGICAL DIVERSITY

Estimated Number of Fungal Species

Among eukaryotic organisms, biological diversity is often measured in numbers of species (Wilson, 1988). In attempting to estimate the total number of species of fungi, both described and undescribed, Hawksworth (1991, 1993) compared the fungus-vascular plant ratios in geographic areas that had been relatively well-studied. He arrived at an estimate of 1.5 million fungal species, based partly on an estimated ratio of unique fungal species to vascular plant species of 6:1. Even if one uses a conservative estimate such as the 1 to 1 ratio of Farr, et al., (1989), the total number of fungal species on vascular plants would be at least 250,000. However, many fungi occur on substrates for which the plant-derived host taxon cannot be identified, such as soil, litter and rotten wood. Another enormous group of fungi to be considered are those associated with insects, of which estimates range from 500,000 species (Blackwell, pers. comm.) to 1.5 million species (Hywel–Jones, 1993). Other substrates include other fungi, animals, and animal parts such as hair, skin, bone, and the digestive tract of ruminants. Additional evidence to support the worldwide estimate of 1.5

million fungal species is the 50-60% rates of newly described species in some fungal genera monographed during the last twenty years (Hawksworth, 1991) and the information on numbers of species in specific countries or on specific hosts that have been relatively well-studied. For example, based on the currently accepted ratio of fungi to vascular plants, Pascoe (1990) estimated that in Australia there exist 250,000 fungal species. Examining a group of obligate parasites of vascular plants, Hennen and McCain (1993) calculated an estimated 25,000-125,000 species of rust fungi (Uredinales). Hsieh and Goh (1993) reported one species of *Cercospora sensu lato* for every ten species of vascular plants in Taiwan. Thus, in one fungal genus alone, about 25,000 species may exist worldwide. The latter two groups of plant parasites constitute a relatively small portion of the fungi as a whole. In exploring the biosphere, scientists, even mycologists, have only barely begun to recognize the immerse diversity that exists.

Fungal Diversity in a Known Universe—an Inventory of All Fungi in a Finite Area

At present, there is literally no single place in the world where all of the living organisms are known, even from one defined area. Such knowledge is the intended result of an all-taxa biodiversity inventory or ATBI, as proposed by Janzen and Hallwachs (1994). In an all-taxa biodiversity inventory, all or almost all the organisms in a relatively large, biologically diverse, terrestrial ecosystem would be discovered and described as fully as possible during five years of intense study. The wealth of information resulting from such a study would be integrated to determine, for example, interrelationships between organisms, or characterized for potential usefulness to humanity.

An inventory of all the fungal taxa in a defined area would sample the complete range of existing substrates in all stages of development and decay over time. At least two approaches would be used, namely direct, visual detection on the substrate, and indirect isolation in laboratory culture separated from the substrate (Bills, 1995; Hawksworth, et al., 1995; Rossman, 1994). Expected results of an all-taxa biodiversity inventory for fungi are estimated at 10,000-50,000 species of which about 50-80% may be new (Rossman, 1994). The totals are based on the estimated number of species for each major fungal group and agree with estimates derived from sites where intense study of the fungi has been undertaken. The only geographical area in which fungi have been relatively well collected is in the United Kingdom. After 25 years of collecting macroscopic fungi in an area of 185 ha, Hawksworth (1993) reported a total of 2,500 fungal species. This site is much smaller than a proposed ATBI with relatively few vascular plant species and the study included only macroscopic fungi collected in the field. Macroscopic fungi constitute less than half the total number of fungal species estimated to exist in a given studied area (Rossman, 1994).

In an ATBI for fungi, a variety of sampling techniques would be applied to numerous substrates covering a relatively large area. Combining data from visual and systematic sampling of macroscopic fungi with those derived from isolating from diverse substrates and habitats suggest that the total number of 10,000-50,000 species is a reasonable estimate for all fungal species to be encountered in an ATBI (Rossman, 1994).

Diversity in Ability to Exploit Nutrients—Tropical Rainforest Leaf Litter

Abundant fungal diversity exists, even in what would be considered a relatively simple substrate. Bills and Polishook (1994) adapted the particle filtration technique to maximize the diversity of fungal species isolated from a substrate. Whereas traditional isolation techniques select for the reproductive propagules and resting states of fast-growing species that utilize primarily simple carbohydrates, the particle filtration technique favors isolation of those fungi actively growing in the substrate. In their study, fungi were isolated from particle suspensions of tropical rain forest leaf litter. From each 0.1-ml sample 80-145 different fungal species were isolated and four replicates of the same substrate yielded between 300-400 different species of fungi. In a second study (Bills & Polishook, 1995), fungi were isolated from only one vascular plant host, *Heliconia mariae*. From each 0.1-ml suspension of filtered particles of leaves from four plants, they isolated about 50-100 different fungal species. In both of these studies, approximately 40-60% of the fungi isolated were rare or unidentifiable. Rarefaction curves calculated for both studies indicated that, despite the isolation of hundreds of isolates from each sample, the maximum number of fungal species had not yet been obtained (Fig. 1). Clearly the fungi are more diverse than has been recognized, even by mycologists.

Diversity in Ability to Exploit Nutrients—Mycorrhizal Fungi

The association of fungi with the roots of terrestrial plants is of ancient origin and has developed numerous times, contributing to the success of both groups of organisms. As summarized in Molina, et al. (1992), 82% of all plant species examined distributed among 90% of the vascular plant families are known to develop mycorrhizae. The particular fungal species associated with roots of one host may vary with plant age, as well as with environmental conditions. Furthermore, an individual plant may be associated with more than one fungus at a time. The mycorrhizal fungi obtain carbohydrates from the plant host accumulated as fungal-specific sugars, while the fungal hyphae function as extensive root hairs for the plant. In one study up to one-third of the carbohydrates produced by the vascular plant host was consumed by the fungal root associates (Molina, et al., 1992). This high nutritional expense suggests that these root fungi are essential for host survival. Mycorrhizae facilitate the uptake of scarce soil nutrients such as phosphorus, tolerance of soil pollutants, particularly heavy metals and aluminum, and protection against plant diseases (Allen, 1992), and are essential to the long-term survival of vascular plants.

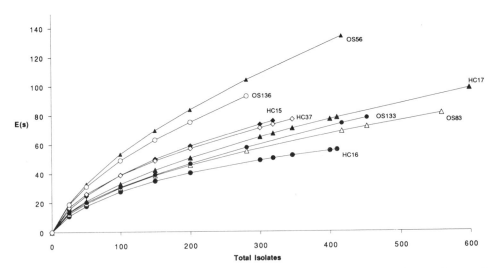

Figure 1. Rarefaction curves for microfungi isolated from mixed leaf litter (OS56-136) and decayed leaves of *Heliconia mariae* from primary forests in Costa Rica (taken from Bills, 1995).

Various kinds of mycorrhizae exist, of which the two most common types are discussed here. Endomycorrhizae, also called vesicular-arbuscular or simply arbuscular mycorrhizae (VAM), are by far the predominant fungus-root association. Arbuscular mycorrhizae are known to occur in over 100,000 species of vascular and non-vascular plants, tending to occur in annual or short-lived plants. Most monocotyledonous plants have VAM as do many dicotyledonous species (Molina, et al., 1992). Certain vascular plant families are remarkable for their lack of VAM, namely the Amaranthaceae and Brassicaceae. On the other hand, the diversity of fungi involved in VAM is limited to about 140 species, all belonging to the Glomales, a group whose phylogenetic relationships are unknown (Morton, 1990). In VAM, the fungal hyphae invade individual host root cells, forming either vesicles or arbuscules within the cells. In the case of arbuscules, the structure gradually dissolves becoming absorbed into the cell cytoplasm, an event correlated with changes in the host cell nucleus. Alternatively, globose fungal vesicles may be formed inside the cell persisting in the soil when the plant cell disintegrates. These characteristic vesicles serve as reproductive propagules that may be transported, for example, by small mammals (McGee & Baczocha, 1994) or merely survive for long periods of time. Known from the fossil record, the existence of mycorrhizae may have contributed significantly to the diversification of vascular land plants (Malloch, et al., 1980; Pirozynski, et al., 1975; Simon, et al., 1993).

In the second kind of mycorrhizae termed ectomycorrhizae, the fungus forms a well-developed tissue enveloping the entire plant root as well as the cells inside the root but does not penetrate the wall of individual cells. Once infected with the fungus, the host root-fungus is modified, developing a morphology characteristic of that particular host-fungus association, often with an extensive branching structure. Ectomycorrhizal associations are known to occur in 7,000 vascular plant species, primarily in perennial, woody plants such as the members of the Fagaceae and Pinaceae. Seemingly anomalous ectomycorrhizal hosts are known, such as one genus in the Poaceae (*Festuca*), and some members of the Ericaceae (Molina, et al., 1992). The fungal species forming ectomycorrhizae are also diverse, ranging from the epigeous Basidiomycetes in the Agaricales (mushrooms, boletes, chanterelles) to hypogeous members of both the Basidiomycetes and Ascomycetes, many of which form multicellular sclerotia contributing to their long term survival. Hypogeous members of the ascomycete order Pezizales, classically separated as the Tuberales, include the highly prized truffles which are obligately ectomycorrhizal with members of the Fagaceae. The peculiar basidiomycetous fungus, *Pisolithus tinctorius*, forms ectomycorrhizae with many host plants belonging to the Fagaceae, Pinaceae, and the genus *Eucalyptus*. Experiments have demonstrated that pine seedlings with this ectomycorrhizal associate are more tolerant of heavy metal soil pollutants and may be able to withstand transplanting into stripmine-spoiled areas. The ubiquitous nature of mycorrhizae suggests that this association has greatly influenced the success of both vascular plants and fungi as evidenced by their present biological diversity.

FUTURE

Fungi have been intimately associated with a diverse range of substrates throughout evolutionary history as symbionts, parasites and saprobes. Mycorrhizal associations were essential for the invasion of terrestrial habitats and are the predominant characteristic of roots of most plant species. The high diversity of fungal species inhabiting a relatively simple substrate, such as rainforest leaf litter suggests that many unique fungal species have evolved to exploit substrates with similar nutritional capacities. As comprehensive projects are undertaken to document the earth's biological diversity, such as in all taxa biodiversity inventories, more evidence of the vast diversity of fungi will become known. Yet, even as the importance and diversity of fungi are increasingly recognized, studies in Europe report a decline in the numbers of ectomycorrhizal and lichenized fungal species (Arnolds, 1988, 1989; Lizon, 1993; Pittam, 1991; Richardson, 1990). Although some species can be maintained *ex situ* in culture collections, many fungi, including those that form mycorrhizae, cannot be maintained axenically in germplasm repositories. Their obligate host associations and specific growth requirements defy long-term preservation when removed from their natural habitat. One must conclude that, in order to preserve diverse fungi essential for the well-being of future generations, it is absolutely crucial to preserve intact ecosystems. Conversely, in order to ensure the long term existence of ecosystems with their myriad of known and unknown biodiversity, urgent measures must be taken to ensure the preservation of the entire spectrum of fungal participants in those ecosystems.

REFERENCES

Allen, M. (ed.) 1992. Mycorrhizal functioning. An integrative plant-fungal process. Chapman-Hall, New York.

Arnolds, E. 1988. The changing macromycete flora in the Netherlands. Trans. Brit. Mycol. Soc. 90:391-406.

Arnolds, E. 1989: A preliminary red data list of macrofungi in the Netherlands. Persoonia 14:77-125.

Baldauf, S.L. and J.D. Palmer. 1993. Animals and fungi are each other's closest relatives: congruent evidence from multiple proteins. Proc. National Acad. Sci. 90:11558-11562.

Berbee, M.L. and J.W. Taylor. 1993. Dating the evolutionary radiations of the true fungi. Canad. J. Bot. 71:1114-1127.

Bills, G.F. 1995. Analyses of microfungal diversity from a user's perspective. Canad. J. Bot. 73, Suppl. 1: 533-541.

Bills, G.F. and J.D. Polishook. 1994. Abundance and diversity of microfungi in leaf litter of a lowland rain forest in Costa Rica. Mycologia 86:187-198.

Bills, G.F. and J.D. Polishook. 1995. Microfungi from decaying leaves of Heliconia mariae. Brenesia:in press.

Bruns, T.D., R. Vilgalys, S.M. Barns, D. Gonzalez, D.S. Hibbett, D.J. Lane, L. Simon, S. Stickel, T.M. Szaro, W.G. Weisburg, and M.L. Sogin. 1992. Evolutionary relationships within the fungi: analyses of nuclear small subunit rRNA sequences. Molec. Phylogenetics and Evolution 1:231-241.

Cavalier–Smith, T. 1987. The origin of fungi and pseudofungi. pp. 339-353. In: Evolutionary biology of the fungi. A.D.M. Rayner, C.M. Brasier, and D. Moore (eds.). Cambridge University, Cambridge.

Dennis, R.L. 1970. A middle pennsylvanian basidiomycete mycelium with clamp connections. Mycologia 62:578-584.

Dick, M.W. 1988. Coevolution in the heterokont fungi (with emphasis on the downy mildews and their angiosperm hosts). In: K.A. Pirozynski and D.L. Hawksworth (eds.) Coevolution of fungi with plants and animals. Academic Press, New York. pp. 31-62.

Farr, D. F., G. F. Bills, G. P. Chamuris, and A. Y. Rossman. 1989. Fungi on Plants and Plant Products in the United States. American Phytopathological Society, St. Paul, Minnesota.

Hass, H., T.N. Taylor,and W. Remy. 1994. Fungi from the lower Devonian Rhynie chert. Amer. J. Bot. 8:29-37.

Hawksworth, D.L. 1991. The fungal dimension of biodiversity: magnitude, significance, and conservation. Mycol. Res. 95:641-655.

Hawksworth, D.L. 1993. The tropical fungal biota: census, pertinence, prophylaxis, and prognosis. In: S. Isaac, J.C. Frankland, R. Watling, and A.J.S. Whalley (eds.) Aspects of Tropical Mycology. Cambridge University Press, Cambridge. pp. 265-293.

Hawksworth, D.L., D.W. Minter, G.C. Kinsey, and P.F. Cannon. 1995. Inventorying a tropical fungal biota: intensive and extensive approaches.

Hennen, J.F. and J.W. McCain. 1993. New species and records of Uredinales from the Neotropics. Mycologia 85:970-986.

Hsieh, W.H. and T.K. Goh. 1993. *Cercospora* and similar fungi from Taiwan. Maw Chang, Taipei.

Hywel–Jones, N.L. 1993. A systematic survey of insect fungi from natural, tropical forest in Thailand. Abstract. In: S. Isaac, J.C. Frankland, R. Watling, and A.J.S. Whalley (eds). Aspects of Tropical Mycology. Cambridge University Press, Cambridge. pp. 295-296.

Janzen, D.H. and W. Hallwachs. 1994. All Taxa Biodiversity Inventory. A Report to the National Science Foundation.

Lizon, P. 1993. Decline of macrofungi in Europe: an overview. Trans. mycol. Soc. R.O.C. 8:21-48.

Malloch, D.W., K.A. Pirozynski, and P.H. Raven. 1980. Ecological and evolutionary significance of mycorrhizal symbioses in vascular plants (A review). Proc. Natl. Acad. Sci. 77:2133-2118.

McGee, P.A. and N. Baczocha. 1994. Sporocarpic Endogonales and Glomales in the scats of Rattus and Perameles. Mycol. Res. 98:246-249.

Molina, R., H. Massicotte, and J.M. Trappe. 1992. Specificity phenomenon in mycorrhizal symbioses: community-ecological consequences and practical implications. pp. 357-423. In: Allen, M.F. (ed) Mycorrhizal Functioning. An Integrative Plant-Fungal Process. Chapman & Hall, New York.

Morton, J.B. 1990. Species and clones of arbuscular mycorrhizal fungi (Glomales, Zygomycetes): their role in macro- and microevolutionary processes. Mycotaxon 37:493-515.

Pascoe, I. G. 1990. History of systematic mycology in Australia. In: P.S. Short (ed.). History of Systematic Botany in Australia. Australian Systematic Botany Society, South Yarra, Australia. pp. 259-264.

Pirozynski, K.A. and Y. Dalpe. 1989. Geological history of the Glomaceae with particular reference to mycorrhizal symbiosis. Symbiosis 7:1-36.

Pirozynski, K.A. and D.W. Malloch. 1975. The origin of land plants: a matter of mycotrophism. BioSystems 6:153-164.

Pittam, S.K. 1991. The rare lichens project, a progress report. Evansia 8:45-47.

Richardson, D.H.S. 1990. Lichens and man. pp. 187-210. In: D.L. Hawksworth (ed.) Frontiers in Mycology. CAB International, Wallingford.

Rossman, A.Y. 1994. A strategy for an all-taxa inventory of fungal biodiversity. In: Peng, C.-I. and C.H. Chou (eds.) Biodiversity and Terrestrial Ecosystems. Inst. Bot., Acad. Sin. Monogr. Ser. 14:169-194.

Sherwood–Pike, M. 1991. Fossils as keys to evolution in fungi. BioSystems 25:121-129.

Sherwood–Pike, M.A. and J. Gray. 1985. Silurian fungal remains: probable records of the Class Ascomycetes. Lethaia 18:1-20.

Simon, L., J. Bousquet, R.C. Levesque, and M. Lalonde. 1993. Origin and diversification of endomycorrhizal fungi and coincidence with vascular land plants. Nature 363:67-69.

Stubblefied, S.P. and T.N. Taylor. 1988. Recent advances in palaeomycology. Tansley Reveiw No. 12. New Phytol. 108:3-25.

Taylor, T.N. and J.F. White, Jr. 1989. Fossil fungi (Endogonaceae) from the Triassic of Antarctica. Amer. J. Bot. 76:389-396.

Taylor, T.N., W. Remy, and H. Hass. 1992. Parasitism in a 400-million-year-old green alga. Nature 357:493-494.

Wainwright, P.O., G. Hinkle, M.L. Sogin, S.K. Stickel. 1993. Monophyletic origins of the metazoa: an evolutionary link with fungi. Science 260:340-341.

Whittaker, R.H. 1969. New concepts of kingdoms of organisms. Science, New York 163:150-160.

Wilson, E.O. (ed.) 1988. Biodiversity. National Acad. Press, Washington, D.C.

6

FUNGAL DIVERSITY AND PHYLOGENY WITH EMPHASIS ON 18S RIBOSOMAL DNA SEQUENCE DIVERGENCE

Junta Sugiyama, Takahiko Nagahama, and Hiromi Nishida

Institute of Molecular and Cellular Biosciences
The University of Tokyo
Yayoi 1-1-1, Bunkyo-ku
Tokyo 113, Japan

INTRODUCTION

"The number of known species of fungi is about 69,000", but species of the fungal world are conservatively estimated to be 1.5 million (Hawksworth, 1991). The fungi are of great consequence agronomically, bioindustrially, medically, and biologically. In spite of their importance, their taxonomic inventory is poor, particularly for the tropical regions. In addition, very little is known about the phylogeny and evolution of fungi and between these and other organisms. As pointed out by Bruns et al. (1991), "their simple and frequently convergent morphology, their lack of a useful fossil record, and their diversity have been major impediments to progress in this field".

To date, phylogenetic speculations and taxonomies for the fungi have been based mainly on analyses of morphological data sets. In the 1980s, development of molecular biological techniques (particularly gene cloning, nucleic acid sequencing, and polymerase chain reaction), proliferation of high performance computers, and improvement of molecular evolutionary analysis programs have extended studies on relatedness, phylogeny, and evolution of organisms, including fungi, at the molecular level. In the early 1990s, such an approach has steered fungal taxonomy towards fungal molecular systematics (Bruns et al., 1991; Hibbett, 1992; Kohn, 1992; Kurtzman, 1992; Sugiyama, 1994). Thus, studies on fungal phylogeny and evolution have entered a new era.

Sexual and asexual reproductive structures have provided important phenotypic characters to measure relatedness and evolutionary affinities among fungi. If they lose these structures, accurate taxonomic assignment is quite difficult. Conversely, nucleic acid characters, as genotypic characters, are ubiquitous and are not dependent on the expression of reproductive structures. Nuclear DNA base composition and nuclear DNA relatedness, as nucleic acid characters, have been used to define species of yeasts (e.g., Kurtzman, 1987; Kurtzman and Phaff, 1987), *Neurospora* (Dutta, 1976), and *Aspergillus* (e.g., Kurtzman et

Microbial Diversity in Time and Space, edited by Colwell et al.
Plenum Press, New York, 1996

al., 1986). However, these molecular characters resolve only to the genetic sibling species level.

Ribosomal RNA sequence comparisons, as a nucleic acid character, offer a means for estimating more distant relationships. Phylogenetic studies of fungi, using 5S rRNA sequence divergence, go back to Walker and Doolittle's work (1982) on several basidiomycetous species. Initially, analyses based on 5S rRNA sequence comparisons improved our understanding of fungal phylogeny and evolution (e.g., Hori and Osawa, 1987; Blanz and Unseld, 1987). Because only 120 nucleotides are available for comparison, however, resolution is limited (Bruns et al., 1991). In recent years, molecular phylogenetic analysis of the fungi shifted to the small (18S) and large (23S to 28S) subunit rRNAs. Phylogenetic analysis among distantly related taxa, using 18S rRNA gene sequences, has contributed to well-resolved and statistically supported conclusions (e.g., Bruns et al., 1991, 1992; Bowman et al., 1992; Berbee and Taylor, 1992; Nishida and Sugiyama, 1993, 1994b; Suh and Sugiyama, 1993, 1994; Swann and Taylor, 1993).

During the past five years, we obtained 18S rRNA gene sequences from the lower to higher fungi to investigate evolutionary relationships among the major groups, and to evaluate existing taxonomic systems and phylogenetic hypotheses that were based mainly on morphological characters. Results of these studies are, in part, described here.

Phylogenetic Hypotheses Concerning the Fungi

Among earlier phylogenetic speculations concerning the fungi and related organisms that have been made during the past 10 years, Cavalier-Smith's theory published in 1987 is noteworthy. He provided a framework for a taxonomic system and phylogeny for the fungal kingdom that was based mainly on cell wall chemistry, biosynthetic pathways for lysine, type of motile cells, cellular ultrastructure, and 5S rRNA sequence divergence. He included only the chytridiomycetes, zygomycetes, ascomycetes, and basidiomycetes in the kingdom Fungi. These four fungal groups are characterized by chitinous cell walls and the alpha-aminoadipic acid (AAA) lysine biosynthetic pathway. The oomycetes, hyphochytrids, labyrinthulids, thraustochytrids, and slime molds, which are cellulosic and have the diaminopimelic acid (DAP) lysine biosynthetic pathway, are excluded from the Fungi. The former four major groups have been accommodated in the pseudofungi and the latter in the slime molds in the kingdom Protozoa (Cavalier-Smith, 1993). In this scheme, he also suggested that Fungi and animals had a common ancestor, a choanociliate protozoan. He concluded that all major eufungal taxa, e.g., the Endomycota, Ascomycota, and Basidiomycota, evolved from the Entomophthorales from a chytridiomycete ancestor by loss of cilia (flagella).

Divergence and Evolutionary Relationships among Lower Fungi

Evidence from 18S rRNA sequence (Föster et al., 1990; Bowman et al., 1992) put an end to the debate as to whether chytrids were true fungi and confirmed that the true fungi, the ascomycetes, the basidiomycetes, the zygomycetes, and the chytridiomycetes, formed a monophyletic group, distinguished from slime molds and the oomycetes. This outline of the true fungi was supported by the presence of chitin in the cell wall (Bartnicki-Garcia, 1970) and the AAA lysine pathway (Vogel, 1964). On the basis of 18S rRNA sequence analysis, Hendriks et al. (1991) indicated that the red algae and the higher fungi did not possess a common ancestor and that their biochemical and ultrastructural morphological similarity were convergent. From maximum likelihood analysis of eukaryotic 18S rRNA sequences (Wainright et al., 1993) it was inferred that animals and the Fungi share a common ancestor, in agreement with Cavalier-Smith's indication (1987) based on selected ultrastructural and biochemical features. Moreover, this result was consistent with results of a comparative study of other molecular sequences (Baldauf and Palmer, 1993).

The true fungi comprise four divisions, i.e., Chytridiomycota, Zygomycota, Ascomycota, and Basidiomycota, and constitute a monophyletic group. In the lower fungi (Chytridiomycota and Zygomycota), primary septation is rare, whereas primary septa predominate in the higher fungi (Ascomycota and Basidiomycota) (e.g., Talbot, 1971). In the view of many mycologists, it is believed that the chytrids are the most primitive fungi because they are zoosporic and gave rise to the zygomycetes. In contrast to the higher fungi, molecular approaches for phylogenetic analysis of the lower fungi are too few (Bruns et al., 1991), mainly because the economic requirements (e. g., the fermentation and biotechnology industries) are low, pure cultures have not been achieved for many species, and even detection of some species is very difficult. However, phylogenetic study of the extensive lower fungal species is indispensable to fill the gaps in knowledge of fungal histories. Molecular phylogenies based on fungal 18S rRNA sequence comparisons (Bruns et al., 1992) showed that the lower fungi, the chytridiomycetes and the zygomycetes, shared basal placements and formed mixed groupings in phylogenetic trees, but the statistical confidences of the branches among the lower fungal lineages were generally low.

As mentioned above, Cavalier-Smith (1987; Fig. 1) suggested that the choanoflagellate was a common ancestor of Fungi and animals and gave rise to the chytridiomycetes. Based on the presence of a procentriole in *Basidiobolus* and the diversity of mitotic ultrastructure in the Entomophthorales, it was inferred that *Basidiobolus* was an ancestral

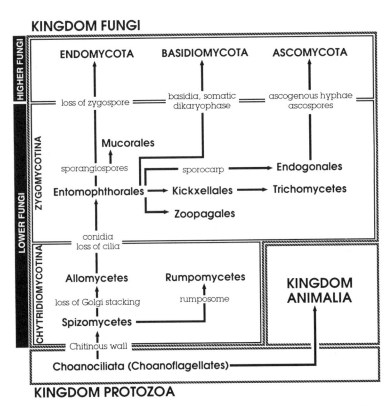

Figure 1. Phylogenetic relationships among the true fungi (Kingdom Fungi). From Cavalier-Smith (1987), with modifications. By permission Cambridge University Press.

zygomycetes, directly evolved from the chytrids, and each of the extant major taxa, the ascomycetes, the basidiomycetes, and the other zygomycetes, diverged radiately from the lineage of the entomophthoralean fungi. However, this evolutionary hypothesis is in disagreement with molecular evidence that the higher fungi form a monophyletic group, distinguished from the lower fungi (e.g., Bruns et al., 1992). Regardless of the conflict, it is remarkable that some information, such as presence of a procentriole in *Basidiobolus* (McKerracher and Heath, 1985) and the ultrastructural morphological and the serological similarities between the Harpellales (Trichomycetes), and the Kickxellales (Moss and Young, 1978) suggest phylogenetic relationships among the major groups.

We constructed phylogenetic trees (Fig. 2) from published 18S rRNA sequences for the lower fungi using neighbor-joining and maximum likelihood methods and compared the trees with existing classifications and the proposed hypothesis of fungal evolution. We used the choanoflagellate, *Diaphanoeca grandis*, as an out group, which has been suggested to be the protozoan ancestor of the true fungi (Cavalier-Smith, 1987). Between the neighbor-joining tree and the maximum likelihood tree, there were differences in the phylogenetic

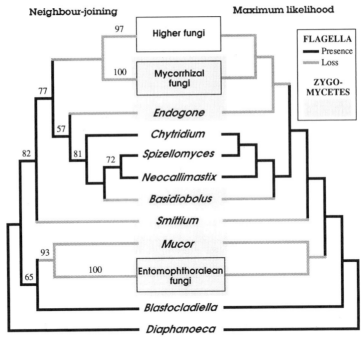

Figure 2. Comparison of branching patterns between neighbour-joining tree and maximum likelihood tree, inferred from 18S rDNA sequence data. Published sequences, for the four entomophthoralean fungi (*Basidiobolus ranarum*, *Conidiobolus coronatus*, *Entomophthora muscae* and *Zoophthora radicans*), the trichomycete Smittium culisetae, the three higher fungi (the two ascomycetous yeasts *Saccharomyces cerevisiae* and *Schizosaccharomyces pombe*, and the basidiomycete *Athelia bombacina*), the four chytrids (*Blastocladiella emersonii*, *Chytridium confervae*, *Neocallimastix* sp. and *Spizellomyces acuminatus*), the endomycorrhizal fungus *Glomus etunicatum*, the zygomycete *Mucor racemosus*, and the choanoflagellate *Diaphanoeca grandis*, were aligned by CLUSTAL W (Thompson et al., 1994). Topology of the left tree was constructed by the neighbor-joining method (Saitou and Nei, 1987), from alignment of the data sets. The percentage of bootstraps (Felsenstein, 1985) was derived from 1,000 resamplings, using CLUSTAL W (Thompson et al., 1994). Values below 50% are not shown. The topology of the right hand tree was constructed using the maximum likelihood program DNAML from PHYLIP 3.5c (Felsenstein, 1994). The lineages where flagella were lost are drawn in gray.

positions of *Blastocladiella*, *Endogone*, and the chytrids, but statistical confidences by bootstrap analysis, supporting the branches of each species, were not high. Neither of the trees supported the fundamental framework that separated the chytridiomycetes from the zygomycetes and estimated that losses of flagella occurred in several lineages during the course of fungal evolution. The concept that some different chytrids lost flagella and gave rise to major groups of filamentous fungi did not correspond with Cavalier–Smith's hypothesis (1987). The presence of zygospores, commonly observed for species in the zygomycetes, may be a convergent character and the definition of zygospores in fungal diversity should be investigated further using ultrastructural data.

Basidiobolus as an evolutionary key fungus, which was considered to be a link between zoosporic and non-zoosporic fungi within the true fungi, joined with the chytrids, Chytridium, Spizellomyces, and Neocallimastix, and formed a cluster with a high level (81%) of bootstrap confidence in the neighbour-joining tree. In the maximum likelihood tree, in spite of changes in branching orders among the four species, their monophyletic grouping was supported. Higher bootstrapping value for the grouping was indicated from the other neighbour-joining tree (Nagahama et al., 1995). Evolutionary consideration between *Basidiobolus* and the zoosporic fungi by the presence of a procentriole was also confirmed from topologies of the phylogenetic tree, based on fungal 18S rRNA sequences, but these results revealed that the entomophthoralean fungi, including *Basidiobolus*, are not monophyletic. Blastocladiella assigned to the Allomycetes, which has been regarded as chytridiomycete ancestors for the zygomycetes, did not show a close relationship to *Basidiobolus*. The other members in the lower fungi, the three mycorrhizal fungi, Endogone, and Smittium, did not strongly join with the species included in the tree.

The phylogenetic scheme of fungi, based on dual combinations of major wall polysaccharides, indicates that the Zygomycetes could be regarded as the chitin-chitosan group (Bartinicki–Garcia, 1970, 1987), with the exceptions of the Entomophthorales (Hoddinott and Olsen, 1972) and the Glomineae in the Glomales (Gianinazzi–Pearson et al., 1994). The heterogeneity of the cell wall polysaccharide suggests their phylogenetic diversity in the zygomycetes, as was supported by topologies of the lower fungal tree. The position of *Basidiobolus* in the tree, diverged from the other entomophthoralean fungi, was supported by the specific carbohydrate compositions of the cell wall (Hoddinott and Olsen, 1972). Tree topologies of the other three species of Entomophthora, Zoophthora, and Conidiobolus for the Entomophthorales were consistent with their taxonomic system (Humber, 1989).

Ultrastructural observations (Saikawa, 1989) of a tardigrade parasite, *Ballocephala verrucospora* (Entomophthorales, Zygomycetes) showed that the entomophthoralean fungi had bifurcately flared septa, similar to those observed in members of the Harpellales and Asellariales (Trichomycetes), and Dimargaritales and Kickxellales (Zygomycetes). The morphological continuities may indicate phyletic relationships among zygomycetes producing a bifurcate septum and a polyphyletic origin for the entomophthoralean fungi.

By overcoming the ambiguity of phylogenetic placement of the entomophthoralean fungi, arising from poor morphological characters and difficult detection, it may be possible to recognize new linkages among major groups of the lower fungi. To establish a taxonomic system of the lower fungi which reflects their evolutionary relationships, further studies of their molecular phylogenetics, chemotaxonomy, and micromorphology are necessary.

Divergence and Evolutionary Relationships among Higher Fungi

Recently, Nishida and Sugiyama (1994b) detected a major new lineage of archiascomycetes within the Ascomycota, based primarily on 18S rRNA gene sequence analysis. This new lineage is a diverse assemblage containing parasitic and saprobic fungi, *Taphrina wiesneri*, *T. populina*, *Protomyces inouyei*, *P. lactucae-debilis*, *Saitoella complicata*, *Schi-*

zosaccharomyces pombe, and *Pneumocystis carinii*. Their mycological profiles are briefly provided, as follows.

Taphrina. The phylogenetic position of *Taphrina* (Taphrinales) has attracted the attention of mycologists interested in evolution of higher fungi. For example, Savile (1955, 1968) has proposed that *Taphrina* represented an early divergence within the higher fungi. The life cycle of *Taphrina deformans*, the peach leaf curl fungus, in the order Taphrinales of the Ascomycota is unique (Kramer, 1987). *Taphrina* species are dimorphic plant parasites, forming mycelium and asci in their parasitic phase and budding yeast cells in their saprobic phase. They lack ascomata. During the parasitic phase, they develop dikaryotic mycelial cells within plant tissues. After meiosis, haploid ascospores develop into budding yeasts that appear as pink colonies on artificial media. *Taphrina wiesneri* (=*T. cerasi*) attacks Japanese cherry trees (*Cerasus yedoensis*), causing "witches' brooms" (Tubaki, 1978).

Protomyces. *Protomyces inouyei* is parasitic, causing galls on leaves and fruit of *Youngia japonica* in Japan and forming in the tissues of its host large, round, thick-walled resting chlamydospores, a result of enlargement of segments of the mycelium (Tubaki, 1957). The phylogenetic position of the genus *Protomyces* (Protomycetales) has been proposed by mycologists, such as Reddy and Kramer (1975), based mainly on morphological data, but it remains uncertain.

Saitoella. *Saitoella*, for which only the type species, *S. complicata*, has been described, represents an anamorphic, saprobic yeast (Goto et al., 1987). The root of *S. complicata* is comprised of two Himalayan yeast isolates, identified as *Rhodotorula glutinis* (Fres.) Harrison by Goto and Sugiyama (1970). The results of chemotaxonomic and ultrastructural studies of species of *Rhodotorula* and its teleomorphic genus, *Rhodosporidium*, led to the proposal of a new genus, *Saitoella* (for a review on the *S. complicata* study, see Sugiyama et al., 1993). Goto et al. (1987) supposed a close relationship between the anamorphic genus *Saitoella* and Taphrinales of the Ascomycota, on the basis of the following points. *Saitoella complicata* and *T. wiesneri* share some characteristics with the ascomycetes and basidiomycetes (Table 1). In both species, the negative diazonium blue b (DBB) reaction and negative extracellular DNase activity resemble characteristics of ascomycetous yeasts, whereas the positive urease activity and major ubiquinone system Q-10 resemble those of basidiomycetous yeasts. As for the mode of budding, these two species are of the enteroblastic type, typical of the basidiomycetous yeasts.

Schizosaccharomyces. The type species *Schizosaccharomyces pombe* is saprobic and characterized by exclusive fission-type vegetative reproduction and the Q-10 system as the

Table 1. Characteristics of *Taphrina wiesneri*, *Saitoella complicata*, and the yeasts

Characters	*Taphrina wiesneri*	*Saitoella complicata*	Ascomycetous yeasts	Basidiomycetous yeasts
Meiosporangium	ascus	lacking	ascus	basidium
Cell wall	?	two-layered	two-layered	multi-layered
Budding	enteroblastic	enteroblastic	holoblastic	enteroblastic
DBB color test	negative	negative	negative	positive
Urease test	positive	positive	negative	positive
DNase test	negative	negative	negative	positive
Major ubiquinone	Q-10	Q-10	Q-6~9	Q-8~10
GC content (mol %)	49.5	51.6	<50	>50

Exceptional data were omitted.

major ubiquinone. It was traditionally included with the ascomycetous yeasts of the Endomy-cetales, but recently, Kurtzman placed this genus in a separate order, the Schizosaccharomy-cetales (Kurtzman, 1993).

Pneumocystis. *Pneumocystis carinii* is a principal causal agent of pneumonia in patients with AIDS (acquired immunodeficiency syndrome). This organism was considered to be a protozoan for many years; however, Edman et al. (1988) showed *P. carinii* to be a fungus, based on 18S rRNA sequence comparisons. On the other hand, the 5S rRNA sequence data suggested *P. carinii* to be closely related to zygomycetes, but not within the higher fungi, nor with other protozoa (Watanabe et al., 1989).

Nishida and Sugiyama (1993) presented molecular phylogeny results that clearly indicate two divisions, Ascomycota and Basidiomycota, among the higher fungi, with strong bootstrap support. Both divisions appear to be monophyletic as already suggested by 18S rDNA sequence comparisons (e.g., Bruns et al., 1992; Berbee and Taylor, 1993; Nishida and Sugiyama, 1993, 1994a, b). The phylogenetic tree (Nishida and Sugiyama, 1993; cf., Nishida et al., 1993) showed that the ascomycetes are composed of three major lineages, one major lineage, with 86% bootstrap confidence, containing *Taphrina wiesneri*, *Saitoella complicata*, and the fission yeast *Schizosaccharomyces pombe*, distinguished from the ascomycetous yeasts (the hemiascomycete lineage: *Kluveromyces lactis*, *Saccharomyces cerevisiae*, and *Candida albicans*) and the filamentous yeasts (the euascomycete lineage: *Neurospora crassa* and *Podospora anserina*). The phylogenetic tree has also suggested that the ascomycetes are not ancestral to the basidiomycetes. As shown in Fig. 3, *Taphrina wiesneri*, *T. populina*, *Protomyces inouyei*, *P. lactucae-debilis*, and *Saitoella complicata*, the 18S sequences of which were determined by us, *Schizosaccharomyces pombe*, and *Pneumocystis carinii* formed a major new lineage within the Ascomycota. This major lineage, i.e., the archias-comycetes, newly described by Nishida and Sugiyama (1994b), diverged prior to separation of other two major lineages. The archiascomycetes correspond to the basal ascomycetes (Berbee and Taylor, 1993) or the early ascomycetes (Taylor et al., 1994). In Fig. 3, the ascomycetous yeasts and filamentous yeasts are shown to be monophyletic, whereas the archiascomycetes may not be monophyletic because of comparatively low bootstrap support (ca. 71%). At the moment, common characters of these eight species, which define the archiascomycetes, are limited. The archiascomycetes are characterized by having a hyphal or yeast-like assimilative state, a sexually ascogenous (but lacking ascogenous hyphae) reproductive state, and an asexually budding or fission process, and by having the major ubiquinone system Q-10 (but lacking ubiquinone data for *Pneumocyctis carinii*).

The species of the *Taphrina* / *Protomyces* branch, which asexually produce the anamorphic yeast state, are distant from both the ascomycetous yeasts and filamentous ascomycetes. The bootstrap analysis indicated 100% support for monophyly of this branch.

Fig. 3 shows the phylogenetic placement of the anamorphic yeast *Saitoella compli-cata*. Our molecular phylogenetic data suggest that presumably Saitoella diverged from a common ancestor related to *Taphrina* and *Protomyces*. Such a phylogenetic speculation is backed by both phenotypic and genotypic data (Table 1; Goto et al., 1987).

The rDNA data indicate that *Schizosaccharomyces pombe* (Endomycetales) and *Pneumocystis carinii* (*Incertae sedis*), both having a fission-type of asexual reproduction, are phylogenetically separate from the ascomycetous yeasts. However, the branching order of *S. pombe* and *P. carinii* is unresolved, because placement among the archiascomycetes is statistically weak (Nishida and Sugiyama, 1994a, b). Possible phylogenetic relationships between these two species are fascinating. Recently, detailed discussion of this point was provided in detail by Taylor et al. (1994).

"Strain IFO 32408", shown in Fig. 3, is *Mixia osmundae*, listed as *Taphrina osmunda* in the IFO List of Cultures, Ninth Edition, p. 426, 1992. Originally this species was described by T. Nishida (1911) as *Taphrina osmundae*, parasitic on *Osmunda japonica* in Japan.

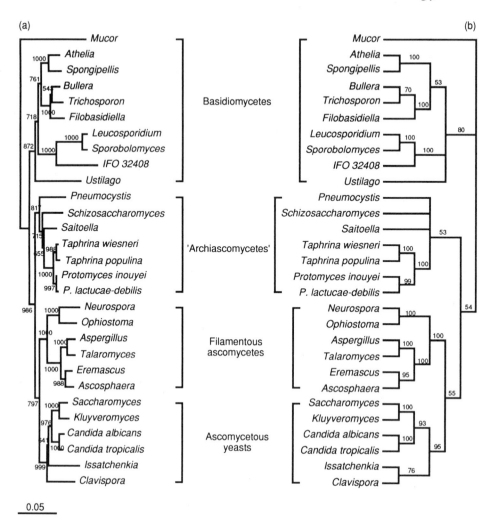

Figure 3. Phylogenetic analyses of higher fungal 18S rDNA sequences. Source information for 18S rDNA sequences has been given in Nishida and Sugiyama (1994b), and Nishida et al. (1995). (a) Neighbor-joining tree. The bootstrap procedure sampled 1,000 times, with replacement by CLUSTAL V (Higgins et al., 1992). The bar indicates 5% difference. (b) Majority-rule consensus tree constructed from a cladistic analysis using 100 times resampling bootstrapping in PAUP ver. 3.0 (Swofford, 1989). The numbers represent frequency of occurrence for that clade, and branching construction was limited to frequencies of occurrence more than 50%.

Kramer (1958, 1987) proposed a new genus, *Mixia*, typified by *Taphrina osmundae*, and subsequently placed in a new family Mixiaceae (Protomycetales). To date, almost all mycologists have accepted *Mixia* as a member of the ascomycetes, related to the species of the Protomycetales and Taphrinales. The molecular-based tree (Fig. 3) indicates that *M. osmundae* and the basidiomycetous yeasts *Leucosporidium* and *Sporobolomyces* group cluster in 100% of the bootstrap replicates. Subsequently, Nishida et al. (1995) clarified this phylogenetic placement, with supportive evidence using both molecular and morphological characters. Obviously, *M. osmundae* is a member of the basidiomycetes and is placed within the simple septate basidiomycete lineage, as defined by Swann and Taylor, (1993), and Suh and Sugiyama (1994).

Finally, a brief note on basidiomycete phylogeny is provided below. The origin of the basidiomycetes is still ambiguous and controversial. However, the 18S rDNA sequence comparisons (Swann and Taylor, 1993; Suh and Sugiyama, 1994; cf., Fig. 3) showed that the basidiomycetes comprise three major lineages; Ustilaginales smuts (or smut fungi); simple septate basidiomycetes, including the teliospore-forming yeasts; and hymenomycetes.

CONCLUSIONS

Molecular phylogenetic analysis, using 18S rDNA sequence divergence, shows that a monophyletic kingdom for the Fungi contains the chytridiomycetes and the zygomycetes as lower fungi, and the ascomycetes and basidiomycetes as higher fungi. It suggests great phylogenetic divergence among the chytrids and the entomophthoralean fungi of the lower fungi and loss of flagella within several lineages of the lower fungi. On the other hand, the molecular and morphological data sets clearly indicate existence of two monophyletic Divisions, Ascomycota and Basidiomycota. The former comprises the archiascomycetes, as a new concept, the hemiascomycetes (ascomycetous yeasts), and the euascomycetes, whereas the latter contains the Ustilaginales smut (or smut fungi), the simple septate basidiomycetes, including most of the basidiomycetous yeasts, and the hymenomycetes. Analyses of more taxa within these lineages and more sequence data are required to elucidate evolutionary relationships among the fungi, from the lower to higher fungi, in light of the extensive fungal species diversity that has been revealed to date.

REFERENCES

Baldauf, S. L. and Palmer, J. D., 1993, Animals and fungi are each other's closest relatives: Congruent evidence from multiple proteins, Proc. Natl. Acad. USA 90: 11558-11562.

Bartnicki–Garcia, S., 1970, Cell wall composition and other biochemical markers in fungal phylogeny. In: Harborne JB (ed) Phytochemical Phylogeny, pp. 81-103. Academic Press, London, United Kingdom.

Bartnicki–Garcia, S., 1987, The cell wall: a crucial structure in fungal evolution. In: Rayner ADM, Brasier CM, and Moore D (ed) Evolutionary Biology of the Fungi, pp. 389-403. Cambridge University Press, Cambridge.

Berbee, M. L., and Taylor, J. W., 1992, Two ascomycete classes based on fruiting-body characters and ribosomal DNA sequences, Mol. Biol. Evol. 9: 278-284.

Berbee, M. L. and Taylor, J. W., 1993, Dating the evolutionary radiations of the true fungi, Can. J. Bot. 71: 1114-1127.

Blanz, P. A. and Unseld, M., 1987, Ribosomal RNA as a taxonomic tool in mycology. In: de Hoog GS, Smith MTh, and Weijman ACM (ed) The Expanding Realm of Yeast-like Fungi, pp. 247-258. Elsevier Science Publishers, Amsterdam.

Bowman, B. H., Taylor, J. W., Brownlee, A. G., Lee J., Lu, S.-D., and White, T. J., 1992, Molecular evolution of the fungi: Relationship of the Basidiomycetes, Ascomycetes, and Chytridiomycetes, Mol. Biol. Evol. 9: 285-296.

Bruns, T. D., Vilgalys, R., Barns, S. M., Gonzalez, D., Hibbett, D. S., Lane, D. J., Simon, L., Stickel, S., Szaro, T. M., Weisburg, W. G., and Sogin, M. L., 1992, Evolutionary relationships within the Fungi: Analyses of nuclear small subunit rRNA sequences, Mol. Phylogenet. Evol. 1: 231-241.

Bruns, T. D., White, T. J., and Taylor, J. W., 1991, Fungal molecular systematics, Ann. Rev. Ecol. Syst. 22: 525-564.

Cavalier–Smith, T., 1987, The origin of Fungi and pseudofungi. In: Rayner ADM, Brasier CM, and Moore D (ed) Evolutionary Biology of the Fungi, pp. 339-353. Cambridge Univ. Press, Cambridge.

Cavalier–Smith, T., 1993, Kingdom Protozoa and its 18 phyla, Microbiol. Rev. 57: 953-994.

Dutta, S. K., 1976, DNA homologies among heterothallic species of Neurospora, Mycologia 68: 388-401.

Edman, J. C., Kovacs, J. A., Masur, H., Santi, D. V., Elwood, H. J., and Sogin, M. L., 1988, Ribosomal RNA sequence shows Pneumocystis carinii to be a member of the fungi, Nature 334: 519-522.

Felsenstein, J., 1985, Confidence limits on phylogenies: an approach using the bootstrap, Evolution 39: 783-791.

Felsenstein, J., 1994, PHYLIP - phylogenetic inference package, version 3.5c, Computer programs distributed by the author, Department of Genetics, University of Washington, Seattle.

Föster, H., Coffey, M. D., Elwood, H., and Sogin, M. L., 1990, Sequence analysis of the small subunit ribosomal RNAs of three zoosporic fungi and implications for fungal evolution, Mycologia 82: 306-312.

Gianinazzi-Pearson, V., Lemoine, M-C., Arnould, C., Gollotte, A., and Morton, J. B., 1994, Localization of b (1Å®3) glucans in spore and hyphal walls of fungi in the Glomales, Mycologia 86: 478-485.

Goto, S. and Sugiyama, J., 1970, Studies on Himalayan yeasts and molds, IV, several asporogenous yeasts including two new taxa of Cryptococcus, Can. J. Bot. 48: 2097-2101.

Goto, S., Sugiyama, J., Hamamoto, M., and Komagata, K., 1987, Saitoella, a new anamorph genus in the Cryptococcaceae to accommodate two Himalayan yeast isolates formerly identified Rhodotorula glutinis, J. Gen. Appl. Microbiol. 33: 75-85.

Hawksworth, D. L., 1991, The fungal dimension of biodiversity: magnitude, significance, and conservation, Mycol. Res. 95: 641-655.

Hendriks, L., De Baere, R., Van de Peer, Y., Neefs, J., Goris, A., and De Wachter, R., 1991, The evolutionary position of the rhodophyte Porphyra umbilicalis and the basidiomycete Leucosporidium scottii among other eukaryotes as deduced from complete sequences of small ribosomal subunit RNA, J. Mol. Evol. 32: 167-177.

Hibbett, D. S., 1992, Ribosomal RNA and fungal systematics, Trans. Mycol. Soc. Japan 33: 533-556.

Higgins, D. G., Bleasby, A. J., and Fuchs, R., 1992, CLUSTAL V: Improved software for multiple sequence alignment, Cabios 8: 189-191.

Hoddinott, J., and Olsen, O. A., 1972, A study of the carbohydrates in the cell walls of some species of the Entomophthorales, Can. J. Bot. 50: 1675-1679.

Hori, H. and Osawa, S., 1987, Origin and evolution of organisms as deduced from 5S ribosomal RNA sequences, Mol. Biol. Evol. 4: 445-472.

Humber, R. A., 1989, Synopsis of a revised classification for the Entomophthorales (Zygomycotina), Mycotaxon 36: 441-460.

Kohn, L. M., 1992, Developing new characters for fungal systematics: An experimental approach for determining the rank of resolution, Mycologia 84: 139-153.

Kramer, C. L., 1958, A new genus in the Protomycetaceae, Mycologia 50: 916-926.

Kramer, C. L., 1987, The Taphrinales. In: de Hoog GS, Smith MTh, and Weijman ACM (ed) The Expanding Realm of Yeast-like Fungi, pp.151-166. Elsevier Science Publishers, Amsterdam.

Kurtzman, C. P., 1987, Prediction of biological relatedness among yeasts from comparisons of nuclear DNA complimentarity. In: de Hoog GS, Smith MTh, and Weijman ACM (ed) The Expanding Realm of Yeast-like Fungi, pp. 459-468. Elsevier Science Publishers, Amsterdam.

Kurtzman, C. P., 1992, rRNA sequence comparisons for assessing phylogenetic relationships among yeasts, Int. J. Syst. Bacteriol. 42: 1-6.

Kurtzman, C. P., 1993, Systematics of the ascomycetous yeasts assessed from ribosomal RNA sequence divergence, Antonie van Leeuwenhoek 63: 165-174.

Kurtzman, C. P. and Phaff, H. J., 1987, Molecular taxonomy. In: Rose AH & Harrison JS (ed) The Yeasts Vol. 1, pp. 63-94. Academic Press, London.

Kurtzman, C. P., Smiley, M. J., Robnett, C. J., and Wicklow, D. T., 1986, DNA relatedness among wild and domesticated species in the Aspergillus flavus group, Mycologia 78: 955-959.

McKerracher, L. J., and Heath, I. B., 1985, The structure and cycle of the nucleus-associated organelle in two species of *Basidiobolus*, Mycologia 77: 412-417.

Moss, S. T., and Young, T. W. K., 1978, Phyletic considerations of the Harpellales and Asellariales (Trichomycetes, Zygomycotina) and the Kickxellales (Zygomycetes, Zygomycotina), Mycologia 70: 944-963.

Nagahama, T., Sato, H., Shimazu, M., and Sugiyama, J., 1995, Phylogenetic divergence of the entomophthoralean fungi: evidence from nuclear 18S ribosomal RNA sequence, Mycologia 87: 203-209.

Nishida, H. and Sugiyama, J., 1993, Phylogenetic relationships among Taphrina, Saitoella, and other higher fungi, Mol. Biol. Evol. 10: 431-436.

Nishida, H. and Sugiyama, J., 1994a, Phylogeny and molecular evolution among higher fungi, Nippon Nôgeikagaku Kaishi 68: 54-57. (In Japanese)

Nishida, H. and Sugiyama, J., 1994b, Archiascomycetes: detection of a major new lineage within the Ascomycota, Mycoscience 35: 361-366.

Nishida, H., Blanz, P. A., and Sugiyama, J., 1993, The higher fungus Protomyces inouyei has two group I introns in the 18S rRNA gene, J. Mol. Evol. 37: 25-28.

Nishida, H., Ando, K., Ando, Y., Hirata, A., and Sugiyama, J., 1995, Mixia osmundae: transfer from the Ascomycota to the Basidiomycota based on evidence from molecular and morphology, Can. J. Bot. 73 (Suppl. 1): S 660-S 661.

Nishida, T., 1911, A contribution to the monogragh of the parasitic Exoascaceae of Japan. In: Collection of Botanical Papers Presented to Prof. Dr. Kingo Miyabe on the Occasion of the Twenty-fifth Anniversary of His Academic Service, pp. 157-204 (in Japanese), pp. 205-212 (English summary). Rokumeik-wan, Tokyo.

Reddy, M. S. and Kramer, C. L., 1975, A taxonomic revision of the Protomycetales, Mycotaxon 3: 1-50.

Saikawa, M., 1989, Ultrastructure of the septum in Ballocephala verrucospora (Entomophthorales, Zygomycetes), Can. J. Bot. 67: 2484-2488.

Saitou, N. and Nei, M., 1987, The neighbour-joining method: a new method for reconstructing phylogenetic trees, Mol. Biol. Evol. 4: 406-425.

Savile, D. B. O., 1955, A phylogeny of the Basidiomycetes, Can. J. Bot. 33: 60-104.

Savile, D. B. O., 1968, Possible interrelationships between fungal groups. In: Ainsworth GC and Susmann AS (ed) The Fungi, An Advanced Treatise, pp. 649-675. Academic Press, New York.

Sugiyama, J., 1994, Fungal molecular systematics: Towards a phylogenetic classification for the fungi, Nippon Nôgeikagaku Kaishi 68: 48-53. (In Japanese)

Sugiyama, J., Nishida, H., and Suh, S.-O., 1993, The paradigm of fungal diagnoses and descriptions in the era of molecular systematics: Saitoella complicata as an example. In: Reynolds DR and Taylor JW (ed) The Fungal Holomorph: Mitotic, Meiotic and Pleomorphic Speciation in Fungal Systematics, pp. 261-269. CAB International, Wallingford, UK.

Suh, S.-O. and Sugiyama, J., 1993, Phylogeny among the basidiomycetous yeasts inferred from small subunit ribosomal DNA sequence, J. Gen. Microbiol. 139: 1595-1598.

Suh, S.-O. and Sugiyama, J., 1994, Phylogenetic placement of the basidiomycetous yeasts Kondoa malvinella and Rhodosporidium dacryoidum, and the anamorphic yeast Sympodiomycopsis paphipedili by means of 18S rRNA gene sequence analysis, Mycoscience 35: 367-375.

Swann, E. C. and Taylor, J. W., 1993, Higher taxa of basidiomycetes: An 18S rRNA gene perspective, Mycologia 85: 923-936.

Swofford, D. L., 1989, PAUP: phylogenetic analysis using parsimony, version 3.0 Illinois Natural History Survey, Campaign, Illinois.

Talbot, P. H. B., 1971, Principal of fungal taxonomy. The Macmillan Press, London. 274 pp.

Taylor, J. W., Swann, E. C., and Berbee, M L., 1994, Molecular evolution of ascomycete fungi: Phylogeny and conflict. In: Hawksworth , D. L. (ed) Ascomycete Systematics: Problems and Perspectives in the Nineties, pp. 201-212. Plenum Press, New York.

Thompson, J. D., Higgins, D. G., and Gibson, T. J., 1994, CLUSTAL W: improving the sensitivity of progressive multiple sequence alignment through sequence weighting, position specific gap penalties and weight matrix choice, Nucl. Acid Res. 22: 4673-4680.

Tubaki, K., 1957, Biological and cultural studies of three species of Protomyces, Mycologia 49: 44-54.

Tubaki, K., 1978, Taphrina wiesneri (Rathay) Mix. In: Udagawa S, Tubaki K et al. (ed) Kinrui-zukan (Compendium of fungi), Part 1, pp. 329-330. Kodansha, Tokyo.

Vogel, H. J., 1964, Distribution of lysine pathways among fungi: evolutionary implication, Am. Nat. 98: 435-446.

Wainright, P. O., Hinkle, G., Sogin, M. L., and Stickel, S. K., 1993, Monophyletic origins of the Metazoa: an evolutionary link with Fungi, Science 260: 340-342.

Walker, W. F. and Doolittle, W. F., 1982, Redividing the basidiomycetes on the basis of 5S rRNA sequences, Nature 299: 723-724.

Watanabe, J., Hori, H., Tanabe, K., and Nakamura, Y., 1989, Phylogenetic association of Pneumocystis carinii with the 'Rhizopoda / Myxomycota / Zygomycota group' indicated by comparison of 5S ribosomal RNA sequences, Mol. Biochem. Parasitol. 32: 163-168.

7

ALGAL DIVERSITY AND EVOLUTION

Hiroshi Oyaizu and Shigeto Ohtsuka

School of Agriculture and Life Science
University of Tokyo
Bunkyo-ku, Tokyo 113, Japan

The earth was born 4.6 billion years ago, as a planet of the solar system. Then, after the chemical evolution of organic compounds ca. 4.0 to 3.5 billion years ago, life was created by accident. Ancient living cells are presumed to have had RNA as their hereditary element (Gilbert, 1986). RNA was substituted by DNA and the present form of living cells (possessing protein-synthesizing ability) is the DNA-based organism. DNA organisms evolved to the presently existing various descendents. Driving forces for the explosive evolution that occurred were environmental changes in the biosphere. The decrease in temperature of the biosphere was one of the most important environmental changes associated with early evolution (Woese, 1981; Woese, 1987). Ancient organisms, living before 3.0 billion years ago, are presumed to have been thermophiles, which required temperatures higher than 70°C. The temperature of the surface of the earth decreased gradually to below 70°C, after 3.0 billion years ago (Trakes, 1979), and thermophilic organisms were, thereafter, found in hot springs or in the hydrothermal vents of deep sea volcanos. The second important environmental change in the biosphere was generation of oxygen (Cloud, 1983). Oxygen was produced by cyanobacteria in the sea, and after oxygen saturation of the sea, oxygen was released into the atmosphere. The generation of oxygen selected against the primitive anaerobic bacteria, leading to selection of aerobic bacteria. The primitive forms of anaerobic bacteria are presumed to be archaebacteria (Woese, 1981; Woese, 1987). As the concentration of atmospheric oxygen increased, aerobic eubacteria diversified explosively about 2.0 billion years ago. The eucaryotes were created by endosymbiosis about 2.0 billion years ago, and various multicellular organisms were created up to 0.7 billion years ago (Cloud, 1983).

Phylogenetic relationships between microorganisms could not be elucidated from fossils. Instead, information on macromolecules, namely, the amino acid sequences of proteins or the nucleotide sequences of genes allowed determination of phylogenetic relationships between and among microorganisms. Included in most living organisms, the macromolecule is concluded to be the molecule existing ubiquitously in all organisms. Phylogenetic relationships drawn from the ribosomal small subunit RNA between and among presently living organisms is shown in Fig. 1 (Wheelis, Kandler, et al., 1992; Woese, 1987; Woese, Kandler, et al., 1990). Very similar phylogenetic trees were drawn using other macromolecules, such as F1 ATPase and elongation factor in peptide synthesis (Iwabe, Kuma, et al., 1989).

Microbial Diversity in Time and Space, edited by Colwell et al.
Plenum Press, New York, 1996

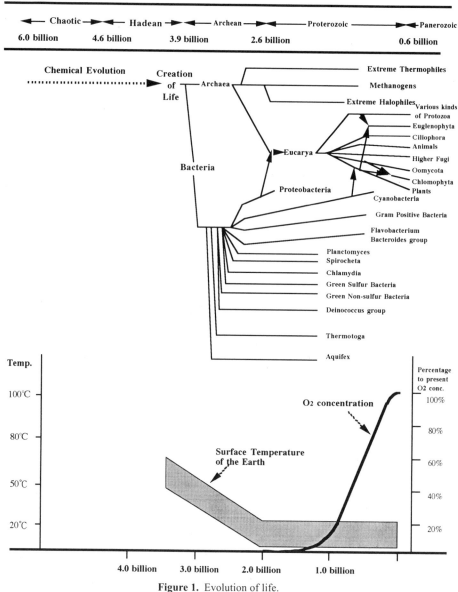

Figure 1. Evolution of life.

Primitive life originating 4.0 to 3.5 billion years ago diversified into three major lineages at a very early stage of evolution (Woese, 1987). The three major lineages have been named domains; Bacteria (Eubacteria); Archaea (Archaebacteria); and Eucarya (Eucaryotes)(Woese, Kandler, et al., 1990). The Bacteria diversified into ca. 12 groups and the Archaea into three groups. The Eucarya comprise animals, plants, fungi, algae, and protozoa. However, this classification does not necessarily reflect phylogenetic relationships. Some of the fungi (the Oomycetes) appear to be related to the brown algae, and some protozoa are related to euglenophytes (Gunderson, Elwood, et al., 1987).

IMPORTANCE OF ENDOSYMBIOSIS FOR THE EVOLUTION OF ALGAE

Endosymbioses played a very important role in the evolution of life (Gray, 1988; Gray, 1989). Energy-producing organs of the eucaryotes, mitochondria, derived from endosymbiosis of a proteobacteria and the organelles of photosynthesis of plants and algae, the chloroplasts, arose from the cyanobacteria.

Algae and plants show heterogeneity in the morphology of their chloroplasts and nuclei, as shown in Fig. 2 (Whatley, John et al., 1979). Eucaryotes with chloroplasts appear on many branches together with organisms without chloroplasts, in molecular phylogenetic trees of cytoplasmic ssu rRNA, and some ,e.g. Euglenophytes, form very deep lines of descent (Bhattacharya and Elwood, 1990; Douglas and Turner, 1991; Eschbach, Wolters, et al., 1991; Gunderson, Elwood, et al., 1987; Hendriks, DeBaere, et al., 1991; Rausch, Larsen, et al., 1989). Therefore, if chloroplasts had a single origin, it can be postulated that the non-photosynthetic eucaryotes lost their chloroplasts independently in evolution. However, since some unicellular algae, such as the chryptomonads, demonstrate anomalous cellular structures (nucleomorph and chloroplast with four membranes) which imply endosymbiosis between two kinds of eucaryotes (Ludwig and Gibbs, 1985; McKerracher and Gibbs, 1982), another hypothesis is that chloroplasts evolved from multiple origins and the latter is considered more attractive (Cavalier-smith, 1986). A molecular phylogenetic study of the cryptomonad algae showed that the chryptomonads were chimaeras of two phylogenetically distinct unicellular eucaryotes (Douglas, Murphy et al., 1991). From accumulated molecular phylogenetic data (Bhattacharya and Elwood, 1990; Douglas, Murphy, et al., 1991; Douglas and Turner, 1991; Eschbach, Wolters, et al., 1991; Gunderson, Elwood, et al., 1987; Hendriks, DeBaere, et al., 1991; Rausch, Larsen, et al., 1989), and electron microscopic analyses (Ludwig and Gibbs, 1985; McKerracher and Gibbs, 1982), the polyphyletic hypothesis involving two step endosymbioses (the first between a cyanobacterium and a phagotrophic eucaryote and the second between the resulting photosynthetic eucaryote and a different phagotrophic eucaryote) seems to be the most probable. To assume a second endosymbiosis is acceptable. However, it is difficult to corroborate how many endosymbioses occurred for the emergence of a special alga, especially when the alga lost its phagocytotic membranes surrounding the chloroplasts.

UNUSUAL PHYLOGENETIC TREES FOUND AMONG THE CHLOROPHYTES

During phylogenetic studies of algal chloroplasts, we determined the ssu rRNA sequences of chloroplasts of representative green algae (Oyaizu, Matsumoto, et al., 1993).

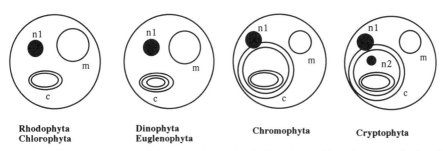

| Rhodophyta | Dinophyta | Chromophyta | Cryptophyta |
| Chlorophyta | Euglenophyta | | |

Figure 2. Chloroplast structure of plants and algae. n1 and n2: nuclei; c: chloroplasts; m: mitochondria.

Phylogenetic relationship between the Chlorophyceae and land plants was inferred by a distance matrix method by using a neighbor-joining algorithm (Saitou and Nei, 1987). To compare phylogenetic trees between chloroplast and cytoplasm is a good approach for discussing the origin of chloroplasts. Thus, phylogenetic trees were constructed employing cytoplasmic and chloroplast ssu rRNAs (Fig. 3). Five *Chlorella* species were found to be extremely divergent in both trees, and the divergence was larger than that found among land plants. According to a widely accepted hypothesis, the class Chlorophyceae and land plants arose from a single ancestor containing chloroplasts, with chlorophyll a and b and surrounded by two membranes. If this hypothesis is true, the tree depicted by cytoplasm ssu rRNA should basically be similar to that depicted by chloroplast ssu rRNA. However, the tree constructed using cytoplasm ssu rRNA was significantly different from the chloroplast ssu rRNA tree. Bootstrap analysis (100 replicates)(Felsenstein, 1985) for the cytoplasmic tree demonstrated 100% confidence for branching between *Chlamydomonas-Chlorella* and land plants, and the same analysis for the chloroplast tree demonstrated reliable confidence (96%) for branching between *Chlamydomonas* and *Chlorella*-land plants. Huss and Sogin (1990) reported the same topology for the cytoplasm-based tree, and Douglas & Turner (1991) and Hendriks et al. (1991) reported the same topology for the chloroplast tree among *Chlorella*, *Chlamydomonas*, and land plants, based on ssu rRNA sequences.

A possible hypothesis to explain the discrepancy is as follows. Molecular evolution of the genes in the cellular organelles was significantly affected by unknown factors

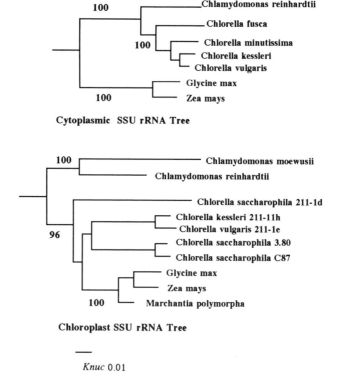

Figure 3. Phylogenetic trees for cytoplasmic ssu rRNA (upper trees) and for chloroplast ssu rRNA (lower tree). The trees were drawn using a neighbor-joining method based on genetic differences, estimated by *Knuc*. *Englena gracilis* was used as the root organism for both trees. The topology of the reconstructed phylogenetic trees were evaluated by the bootstrap sampling (100 replications) method. Confidence limits are shown for each branch.

associated with the host and, thereby, the phylogenetic tree was distorted. Such uncertainty in phylogenetic tree construction occurs when trees are drawn that include very closely related organisms (Saitou and Nei, 1986; Sourdis and Nei, 1988) and organisms for which evolutionary speeds are quite different from each other(Felsenstein, 1978; Woese, 1987). However, such indications were not found in the chloroplast phylogenetic tree. *Chlamydomonas*, *Chlorella*, and land plants are phylogenetically distant and distinct taxa. Furthermore, the evolutionary speed of the three taxa appears to be similar. That is, all branches of the chloroplast tree had similar length from the ancient branching from the root organism (*Englena gracilis*).

The other possible illustration is that the origins of chloroplasts would be different between *Chlamydomonas* and *Chlorella* (and land plants). In this case, the following two theories were hypothesized, placing reliance on the cytoplasm continuity between the class Chlorophyceae and land plants. One is that the ancestors of the Chlorophyceae and land plants formed relatively recently (probably 500 to 600 million years ago, just before the emergence of land plants (Margulis and Schwartz, 1982)) by endosymbioses of several kinds of green algae (procaryotes or eucaryotes) with protists or fungi. In this theory, invasion to the host occurred after diversification of host organisms proceeded to *Chlamydomonas*, *Chlorella*, and land plants. The other theory is that the common ancestor of *Chlamydomonas*, *Chlorella* and land plants had the same chloroplast and, later, after diversification to *Chlamydomonas* and *Chlorella* chloroplasts of *Chlamydomonas* were replaced by other algal endosymbionts, which gave rise to the contemporary *Chlamydomonas* lineage.

The above mentioned hypotheses are somehow abrupt. To recognize the chloroplast tree as an artifact of tree construction may be reasonable. However, a large evolutionary gap was observed between *Chlamydomonas* and land plants in the mitochondrial ssu rRNAs (Gray, Cedergren et al., 1989), and a relatively recent endosymbiotic event involving proteobacteria has been proposed for the evolution of plant mitochondria (Gray, Cedergren et al., 1989). Very likely many complex evolutionary processes took place leading to the evolution of green algae and land plants.

THE IMPORTANCE OF PIGMENT COMPOSITION FOR ALGAL SYSTEMATICS

Pigment (chlorophyll, carotene, xanthophyll, and phycobilin) composition is a very important taxonomic criterion. In particular, the composition of chlorophyll and phycobilin is a very important key character. However, composition does not completely reflect phylogenetic relationships among plants. Euglenophyta and Chlorophyta contain chlorophyll a and b but no phycobilin. However, they are morphologically significantly different, as well as phylogenetically distant from each other. Fig. 4 shows the phylogenetic tree based on chloroplast ssu rRNA. In this tree Euglenophyta (green algae) formed a cluster together with red algae (Rhodophyta (Cyanidium caldarium)), and brown algae (Cryptophyta (Cryptomonas sp.)), and apparently distant from other green algae (Chlorophyta). Fig 5. shows the phylogenetic tree based on cytoplasmic ssu RNA. In the cytoplasmic tree, Euglenophyta is distant from the cluster of red and brown algae.

The compound, 2, 4-Divinylprotochlorophyllide (DVP), had been found only in certain members of the class Prasinophyceae of the division Chlorophyta. Sasa et al. (Sasa, Suda, et al., 1992) studied pigment composition of *Chlamydomonas* parkeae (class Chlorophyceae), and found DVP in the cells of this species. The morphological characteristics of C. parkeae apparently are similar to those of the family Volvocales of the class Chlorophyceae. In the phylogenetic trees (Fig. 4 and 5), C. parkeae was included in the cluster of the

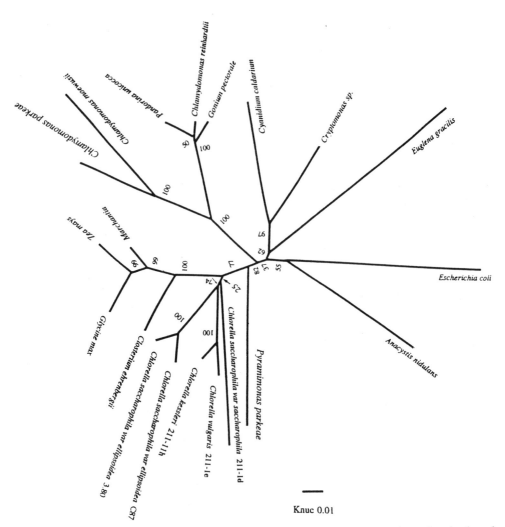

Figure 4. Phylogenetic tree for chloroplast ssu rRNA. The methods of tree construction and evaluation of confidence limits are as in Fig. 3.

family Volvocales. It is concluded that C. parkeae is a very unusual green algae which has Prasinophyceae-like pigment composition (Kim, Oyaizu et al., 1994).

IMPORTANCE OF GENE TRANSFER BETWEEN CELLULAR GENOMES FOR ALGAL EVOLUTION

The chloroplast genomes of plants are circular, and their size ranges from 120 to 150 kb in land plants, from 84 to 2000 kb in green algae, and 100 to 160 kb in non-green algae (Palmer, 1985). The plastid genome of *Epifagus virginiana* is the smallest genome size (Wolfe, Morden, et al., 1992). However, the plant is nonphotosynthetic and parasitic and the plastid is not a functional chloroplast (Wolfe, Morden et al., 1992). As a functional form, the chloroplast genome of *Codium fragile* (chlorophyte) is the smallest size (84 kb) (Manhart,

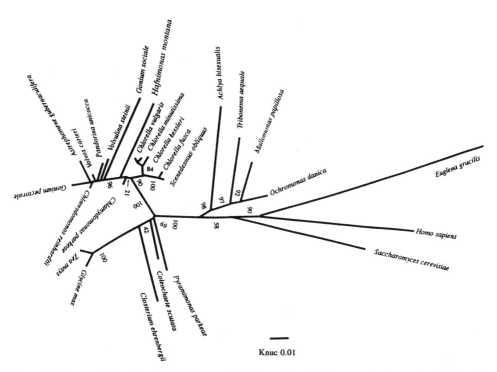

Figure 5. Phylogenetic tree for cytoplasmic ssu rRNA. The methods of tree construction and evaluation of confidence limits are as in Fig. 3.

Kelly, et al., 1989). The largest chloroplast genome is found in some of the chlorophytes (Green, 1976; Padmanabhan and Green, 1978; Tymms and Schweiger, 1985). However, most of the chloroplast genomes described to date fall into a restricted size range of between 120 to 160 kb (Palmer, 1985). The chloroplast genomes contain essential genes for photosynthesis, biosynthesis, and transcription/translation/replication. The chloroplasts appear to require at least 80 kb DNA to accomodate these essential genes (Hiratsuka, Shimada, et al., 1989; Manhart, Kelly, et al., 1989; Ohta, Kawano, et al., 1994; Ohyama, Fukuzawa, et al., 1986; Palmer, 1985; Reith and Munholland, 1993; Shinozaki, Ohme, et al., 1986).

Gene transfer between cellular genomes occurred during various stages of plant evolution (Schuster and Brennicke, 1987). The primitive gene structure of chloroplasts is found in red and brown algae. The chloroplast genomes of non-green algae have many genes which are not found in nuclei of green plants (Table 1). Such a genome structure of chloroplasts of non-green algae shows traces of cyanobacteria-like structure (Douglas, 1992; Pancic, Strotmann, et al., 1992; Scaramuzzi, Stokes, et al., 1992; Valentin, 1993). The genes which are found in chloroplasts in non-green algae and in nuclei of green plants would have transferred from chloroplasts to nuclei during evolution.

Gene transfer also occurs between chloroplasts and mitochondria. In higher plants (maize, mung bean, and *Oenothera*), evidence has been published is for gene transfer, e.g., *trn*P-*trn*W-*pet*G genes (Kubo, Yanai, et al., 1995), *atp*B (Stern and Palmer, 1984), *rbc*L (Lonsdale, Hodge, et al., 1983; Stern and Palmer, 1984), *psb*A (Sederoff, Ronald, et al., 1986), and ribosomal RNA genes (Schuster and Brennicke, 1987) from chloroplasts to mitochondria.

Table 1. Differences in chloroplast gene composition between green algae and non-green algae

Gene	Chlorophyte (Green algae)	Chromophyte and Rhodophyte (Non-green algae)	Reference
rbcS	Nucleus	Chloroplast	(Douglas and Durnford, 1989; Hwang and Tabita, 1991; Reith and Cattolico, 1986; Valentin and Zetsche, 1989)
atpG atpD	Nucleus	Chloroplast	(Pancic, Strotmann, et al., 1992)
secA secY	Not found in chloroplast	Chloroplast	(Douglas, 1992; Pncic, Strotmann, et al., 1992; Valentin, 1993; Valentin and Zetsche, 1989)
hsp70	Not found in chloroplast	Chloroplast	(Craig, Kramer, et al., 1989; Hemmingsen, Woolford, et al., 1988; Pancic, Strotmann, et al., 1992; Scaramuzzi, Stokes, et al., 1992; Valentin, 1993; Valentin and Zetsche, 1989)

Very unusual phylogenetic trees were reported for the large subunit of the ribulose-1,5-bis-phosphate carboxylase/oxygenase (rubisco), rbcL gene. One of the trees reported by Martin et al. (1992) is shown in Fig 6 (a tree drawn by the neighbor joining method). In this tree, green algae and non-green algae formed separate clusters and each cluster contained proteobacteria as sister groups. The green algae were related to cyanobacteria and Chromatium (γ proteobacteria) and the non-green algae were related to *Alcaligenes* (β proteobacteria) and *Rhodobacter* (α proteobacteria). Mitochondria were derived from endosymbiosis of some of the α proteobacteria. Thus, if gene transfer occurred between mitochondria and

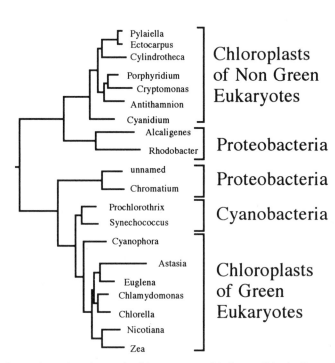

Figure 6. Phylogenetic tree based on nucleotide sequences of rbcL gene (Martin, Somerville, et al., 1992)

chloroplasts, only α proteobacteria should be close relatives to chloroplasts. From these results it is apparent that the origins of the rbcL genes are different between green algae and non-green algae. It is probable that very complex gene transfers occurred between mitochondria, chloroplasts, and other organisms, including β and γ proteobacteria. Furthermore, the genomes of chloroplasts are concluded to be very complex chimeras. Much more information concerning chloroplast genome sequences will be required before the evolution of the chloroplast is fully elucidated.

REFERENCES

Bhattacharya, D. and Elwood, H. J., 1990, Phylogeny of *Gracilaria lemaneiformis* (Rhodophyta) based on sequence analysis of its small subunit ribosomal RNA coding region. J. Phycol. 26, 181-186.

Cavalier–Smith, T., 1986, The kingdom *Chromista*: Origin and systematics. Progress in Phycological Research 4, 309-347.

Cloud, P., 1983, The biosphera. Scientific American 249, 132-144.

Craig, E. A., et al., 1989, *SSC1*, and essential member of the yeast hsp70 multigene family, encodes a mitochondrial protein. Mol. Cell. Biol. 9, 3000-3008.

Douglas, S. E., 1992, A sec Y homologue is found in the plastid genome of *Cryptomonas* φ. FEBS Letters 298, 93-96.

Douglas, S. E. and Durnford, D. G., 1989, The small subunit of ribulose-1,5-bisphosphate carboxylase is plastid-encoded in the chlorophyll c-conteining alga Cryptomonas f. Plant. Mol. Biol. 13, 13-20.

Douglas, S. E., et al., 1991, Cryptomonad algae are evolutionary chimaeras of two phylogenetically distinct unicellular eucaryotes. Nature 350, 148-151.

Douglas, S. E. and Turner, S., 1991, Molecular evidence for the origin of plastids from a cyanobacterium-like ancester. J. Mol. Evol. 33, 267-273.

Eschbach, S., et al., 1991, Primary and secondary structure of the nuclear small subunit ribosomal RNA of the Cryptomonad *Pyrenomonas salina* as inferred from the gene sequence: Evolutionary implications. J. Mol. Evol. 32, 247-252.

Felsenstein, J., 1978, Cases in which parsimony or compatibility methods will be positively misleading. Syst. Zool. 27, 401-410.

Felsenstein, J., 1985, Confidence limits on phylogenies: An approach using the bootstrap. Evolution 39, 783-791.

Gilbert, W., 1986, The RNA world. Nature 319, 618.

Gray, M. W., 1988, Organelle origins and ribosomal RNA. Biochem. Cell Biol. 66, 325-348.

Gray, M. W., 1989, The evolutionary origins of organelles. Trends in Genetics 5, 294-299.

Gray, M. W., et al., 1989, On the evolutionary origin of the plant mitochondrion and its genome. Proc. Natl. Acad. Sci. U.S.A. 86, 2267-2271.

Green, B. R., 1976, Covalently closed minicircular DNA associated with *Acetabularia* chloroplasts. Biochim. Biophys. Acta 447, 156-166.

Gunderson, J. H., et al., 1987, Phylogenetic relationships between chlorophytes, chrysophytes, and oomycetes. Proc. Natl. Acad. Sci. U.S.A. 84, 5823-5827.

Hemmingsen, S. M., et al., 1988, Homologous plant and bacterial proteins chaperone oligomeric protein assembly. Nature 333, 330-334.

Hendriks, L., et al., 1991, The evolutional position of the rhodophyte *Porphyra umbilicalis* and the Basidiomycete *Leucosporidium scottii* among other eukaryotes as deduced from complete sequences of small ribosomal subunit RNA. J. Mol. Evol. 32, 167-177.

Hiratsuka, J., et al., 1989, The complete sequence of the rice (*Oryza sativa*) chloroplast genome: Intermolecular recombination between distinct tRNA genes accounts for a major plastid DNA inversion during the evolution of the cereals. Mol. Gen. Genet. 217, 185-194.

Huss, V. A. R. and Sogin, M. L., 1990, Phylogenetic position of some *Chlorella* species within the Chlorococcales based upon complete small-subunit ribosomal RNA sequences. J. Mol. Evol. 31, 432-442.

Hwang, S. and Tabita, F. R., 1991, Cotranscription, deduced primary structure, and expression of the chloroplast-encoded *rbc*L and *rbc*S genes of the marine diatom *Cylindrotheca* sp. strain N1. J. Biol. Chem. 266, 6271-6279.

Iwabe, N., et al., 1989, Evolutionary relationship of archaebacteria, eubacteria, and eukaryotes inferred from phylogenetic trees of duplicated genes. Proc. Natl. Acad. Sci. USA 86, 9355-9359.

Kim, Y., et al., 1994, Chloroplast small-subunit RNA gene sequence from *Chlamydomonas* parkeae (Chlorophyta): molecular phylogeny of a green alga with a peculiar pigment composition. Eur. J. Phycol. 29, 213-217.

Kubo, T., et al., 1995, The chloroplast *trn*P-*trn*W-*pet*G gene cluster in the mitochondrial genomes of *Beta vulgaris*, *B. trigyna* and *B. webbiana*: evolutionary aspects. Curr. Genet. 27, 285-289.

Lonsdale, D. M., et al., 1983, Maize mitochondrial DNA contains a sequence homologous subunit gene in chloroplast DNA. Cell 34, 1007-1014.

Ludwig, M. and Gibbs, S. P., 1985, DNA is present in the nucleomorph of cryptomonads: Further evidence that the chloroplast evolved from a eukaryotic endosymbiont. Protoplasma 127, 9-20.

Manhart, J. R., et al., 1989, Unusual characteristics of *Codium fragile* chloroplast DNA revealed by physical and gene mapping. Mol. Gen. Genet. 216, 417-421.

Margulis, L. and Schwartz, K. V., 1982, Five Kingdoms. San Francisco, Freeman & Co.

Martin, W., et al., 1992, Molecular phylogenies of plastid origins and algal evolution. J. Mol. Evol. 35, 385-404.

McKerracher, L. and Gibbs, S. P., 1982, Cell and nucleomorph division in the alga *Cryptomonas*. Can. J. Bot. 60, 2440-2452.

Ohta, N., et al., 1994, Physical map of the plastid genome of the unicellular red alga Cyanidium caldarium strain RK-1. Curr. Genet. 26, 136-138.

Ohyama, K., et al., 1986, Chloroplast gene organization deduced from complete sequence of liverwort *Marchantia polymorpha*. Nature 322, 572-574.

Oyaizu, H., et al., 1993, Polyphyletic origins of chlorophyte chloroplasts. J. Gen. Appl. Microbiol. 39, 313-319.

Padmanabhan, U. and Green, B. R., 1978, The kinetic complexity of *Acetabularia* chloroplast DNA. Biochim. Biophys. Acta 521, 67-73.

Palmer, J. D., 1985, Comparative organization of chloroplast genomes. Ann. Rev. Genet. 19, 325-354.

Pancic, P. G., et al., 1992, Chloroplast ATPase genes in the diatom *Odontella sinensis* reflect cyanobacterial characters in structure and arrangement. J. Mol. Biol. 224, 529-536.

Rausch, H., et al., 1989, Phylogenetic relationships of the green alga *Volvox carteri* deduced from small-subunit ribosomal RNA comparisons. J. Mol. Evol. 29, 255-265.

Reith, M. and Cattolico, R. A., 1986, Inverted repeat of *Olisthodiscus luteus* chloroplast DNA contains genes for both subunits of ribulose-1,5-bisphosphate carboxylase and the 32,000-dalton QB protein: Phylogenetic implications. Proc. Natl. Acad. Sci. 83, 8599-8603.

Reith, M. and Munholland, J., 1993, A high-resolution gene map of the chloroplast genome of the red alga *Porphyra purpurea*. Plant Cell 5, 465-475.

Saitou, N. and Nei, M., 1986, The number of nucleotiedes required to determine the branching order of three species, with special reference to the Human-Chimpanzee-Gorilla divergence. J. Mol. Evol. 24, 189-204.

Saitou, N. and Nei, M., 1987, The neighbor-joining method: A new method for reconstructing phylogenetic trees. Mol. Biol. Evol. 4, 406-425.

Sasa, T., et at., 1992, A yellow marine *Chlamydomonas* morphology and pigment composition. Plant Cell Physiol. 33, 527-534.

Scaramuzzi, C. D., et al., 1992, Characterisation of a chloroplast-encoded *sec*Y homologue and *atp*H from a chromophytic alga. FEBS Letters 304, 119-123.

Scaramuzzi, C. D., et al., 1992, Heat shock Hsp70 protein is chloroplast-encoded in the chromophytic alga *Pavlova lutherii*. Plant Mol. Biol. 18, 467-476.

Schuster, W. and Brennicke, A., 1987, Plastid DNA in the mitochondrial genome of Oenothera: intra- and interorganellar rearrangements involving part of the ribosomal cistron. Mol. Gen. Genet. 210,

Sederoff, R. R., et al., 1986, Maize mitochondrial plasmid S-1 sequences share homology with chloroplast gene *psb*A. Genetics 113, 469-482.

Shinozaki, K., et al., 1986, The complete nucleotide sequence of the tobacco chloroplast genome: its gene organization and expression. EMBO J. 5, 2043-2049.

Sourdis, J. and Nei, M., 1988, Relative efficiencies of the maximum parsimony and distance-matrix methods in obtaining the correct phylogenetic tree. Mol. Biol. Evol. 5, 298-311.

Stern, D. B. and Palmer, J. D., 1984, Extensive and widespread homologies between mitochondrial DNA and chloroplast DNA in higher plants. Proc. Natl. Acad. Sci. USA 81, 1946-1950.

Trakes, L. A., 1979, Climates throughout geologic time. Amsterdam, Elsevier Scientific Publishing Co.

Tymms, M. J. and Schweiger, H., 1985, Tandemly repeated nonribosomal DNA sequences in the chloroplast genome of an *Acetabularia mediterranea* strain. Proc. Natl. Acad. Sci. USA 82, 1706-1710.

Valentin, K., 1993, Sec A is plastid-encoded in a red alga: implications for the evolution of plastid genomes and the thylakoid protein import apparatus. Mol. Gen. Genet. 236, 245-250.

Valentin, K. and Zetsche, K., 1989, The genes of both subunits of ribulose-1,5-bisphosphate carboxylase constitute and operon on the plastome of a red alga. Curr. Genet. 16, 203-209.

Whatley, J. M., et al., 1979, From extracellular to intracellular: the establishment of mitochondria and chloroplasts. Proc. R. Soc. Lond. B 204, 165-187.

Wheelis, M. L., et al., 1992, On the nature of global classification. Proc. Natl. Acad. Sci. USA 89, 2930-2934.

Woese, C. R., 1981, Archaebacteria. Scientific American 244, 94-106.

Woese, C. R., 1987, Bacterial evolution. Microbiol. Rev. 51, 221-271.

Woese, C. R., et al., 1990, Towards a natural system of organisms: proposal for the domains Archaea, Bacteria, and Eucarya. Proc. Natl. Acad. Sci. USA 87, 4576-4579.

Wolfe, K. H., et al., 1992, Rapid evolution of the plastid translational apparatus in a nonphotosynthetic plant: loss or accelerated sequence evolution of tRNA and ribosomal protein genes. J. Mol. Evol. 35, 304-317.

8

SYMBIOSIS IN TERMITES

Ikuo Yamaoka

Biological Institute
Faculty of Science
Yamaguchi University
Yamaguchi 753, Japan

INTRODUCTION

Termites are divided into two types, the lower termite, which possesses symbiotic protozoa and several species of bacteria in the intestine and the higher termite, which has no protozoa but several species of symbiotic bacteria in the intestine. The digestive system for cellulose differs in each. In the higher termites, the digestive system is complex, since several problems remain unresolved. In the lower termites, cellulose digestion depends on the intestinal protozoa and symbiotic relationships between the termite and its intestinal protozoa, first published by Cleveland in 1924 (Cleveland, 1924). Recent reports, however, show that the symbiotic relationship in the digestive system involving cellulose is more complex. Yamaoka and Nagatani (1975) reported that the termites themselves produce a cellulase which differs from the cellulase produced by the intestinal protozoa. Since the concept of the cellulose digestion system has changed significantly, the role of intestinal bacteria and their diversity in termites are presented here.

CELLULOSE DIGESTION

The new concept referred to above has been summarized by Yamaoka(1989) and also presented schematically. In brief (Fig.1), native cellulose eaten by the termite is transported to the hindgut, through the foregut and the midgut, without digestion. The cellulose fragments are phagocytized into the phagosome of the intestinal protozoa.

The first degradation step for cellulose is carried out in the phagosome by C1 enzyme. The C1 is a cellulase produced by the protozoa. Secondary digestion is carried out in the phagosome by Cx. Cx is carboxyl-methyl cellulase. Cx is produced by in termite salivary gland, and acts not to degrade the native cellulose, but the linear cellulose chain. The linear

Microbial Diversity in Time and Space, edited by Colwell et al.
Plenum Press, New York, 1996

65

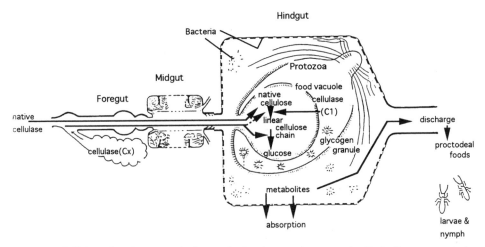

Figure 1. Cellulose digestive system in the termite (worker cast, lower termite, *Reticulitemes speratus*).

cellulose chain is produced by the former process. Cx originates in the salivary gland and is transported into the phagosome along with the native cellulose.

ENVIRONMENT OF THE INTESTINE

Cellulose digestion has been demonstrated by cultivation in vitro of the intestinal protozoa. A conditioned medium is required for cultivation and was prepared as follows.

The hindguts of 10 individual termites were excised and immersed in 10 ml of Trager's solution-U (Trager, 1934). After incubation for three days at 25°C, the solution was centrifuged and 1 ml of supernatant was added to a plastic cell for spectrophotometry and bubbled with nitrogen gas. Freshly prepared intestinal protozoa were inoculated into the

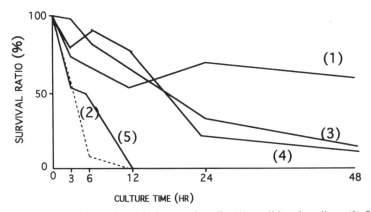

Figure 2. Cultivation of *Trichonympha agilis* in several media. (1) conditioned medium; (2) filtered conditioned medium; (3) bacteria added to the filtered conditioned medium; (4) bacteria (two species) added to centrifuged, conditioned medium; (5) N2 bubbled, conditioned medium.

medium, and surviving protozoa were counted, i.e., after the inoculation was made, zero (0) time, and at the time the survival ratio was calculated, 100 %. Results are shown in Fig.2-(1).

During the first 6 hr, the protozoa increased in number, but during the last 48hr, the survival ratio was very low. Fig.2-(2) was obtained when the conditioned medium was filtered. When the adhering substances on the the filter were resuspended in the medium, the survival ratio improved slightly (Fig.2-(3)).

Recovery, however, was not obtained when only nitrogen gas was bubbled into the medium (Fig.2-(5)).These results suggested that bacteria are required to culture protozoa. Thus, isolation of the intestinal bacteria of the termite was attempted.

Addition of several kinds of intestinal bacteria to the medium yielded improved results, over and above that of bubbling nitrogen gas through the medium (Fig.2-(4)). The bacterial isolates were not identified, but showed high oxygen consumption in solution-U.

OTHER FACTORS INVOLVED IN SURVIVAL

Another observation was the efficiency of the termite saliva. At first we expected that the saliva and midgut contents may contain high cellulase activity. Thus, an homogenate of the salivary gland and the midgut was added to the medium at 6 hr and 12 hr after inoculation of the protozoa. When the saliva was added to the medium, a second cell division was observed at 30hr (Fig.3-(2)).

Similar results were obtained by addition of cellulase prepared from *Aspergillus,* carboxyl-methyl cellulase (CM cellulase) (Fig.3-(3)).

These results suggested that the cellulase produced by the termites themselves is a CM cellulase, i.e.,endoglucanase, i.e., a CM cellulase needed for cellulose digestion and an important factor in inducing the second cell division.

Thus intestinal protozoa appear to play an important role in cellulose digestion, and their cellulose digestive activities are supported by the intestinal bacteria, as well as saliva produced by the termite itself.

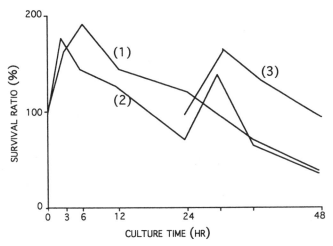

Figure 3. Effect of exogenous cellulase added to the culture medium: (1) control; (2) emended with salivary gland homogenate; (3) emended with CM cellulase from *Aspergillus.*

DIVERSITY AND THE ROLE OF TERMITE INTESTINAL BACTERIA

Cultivation over the long term was the next problem that was investigated. i.e., repeated cell division .This problem is a difficult one but will be resolved, since some important results have been obtained. The first was the discovery (Koga et al.,1992) that all intestinal protozoa of the termite have no mitochondria. Instead, they possess a hydrogenosome. The cytochemical reaction products of the hydrogenase are localized in a special site in the body of *Trichonympha agilis*.Metabolites of the hydrogenosome are hydrogen and bicarbonate, which are not used by the termites, but, in fact, are harmful to them. Messer and Lee (1989) reported that some intestinal protozoa have symbiotic methanogenic bacteria in their cytoplasm, and the bacteria utilize hydrogen and bicarbonate in the hindgut lumen of *Zootermopsis*. The methanogenic bacteria were tested by fluorescence microscopy for Factor 420. A second finding is that similar bacteria were also observed in *Reticulitermes speratus* (Yamaoka and Murakami, 1992; Yamaoka et al.,1993).

Most of the methanogenic bacteria were present also in the inner surface of the hindgut wall. These results suggest that the gut methanogens play an important role in methanogenesis.

Briefly summarized (Fig. 4), is that at least two types of methanogenic bacteria exist in the hindgut, one of them adhered to the cuticle surface and the other a symbiont. Most of the harmful gas, hydrogen, and bicarbonate are converted to methane by the gut methanogens. Symbiotic methanogens have not been shown to have a physiological role until now. Therefore, for long term cultivation, methanogens must be added to the culture medium to increase survival.

16S rRNA SEQUENCES OF METHANOGENIC BACTERIA

Recently, a portion of the 16S ribosomal RNA gene of the methanogenic bacteria present in the gut and in the protozoa was amplified by polymerase chain reaction (PCR) using Methanobacteriales specific primers (Shinzato et al. 1994). Ten individual termites (worker cast) were used. The hindgut was removed and separated into gut tissue and gut contents. The gut tissues contained the gut methanogens and the gut contents contained the

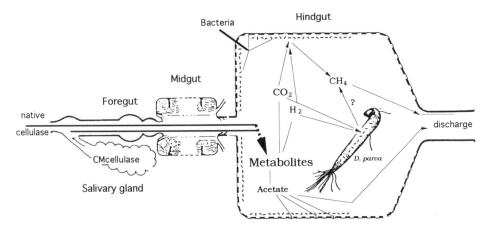

Figure 4. Schematic drawing of the role of the symbiotic bacteria in the termite gut.

Figure 5. 16S rRNA sequence of the bacteria symbiont. The upper and lower lanes represent the symbiont and *M. formicicum*, respectively.

symbiotic bacteria. Each homogenate was frozen and thawed 5 times and incubated in protein K digestive medium at 57°C for overnight. The primers were designed for four species of the Methanobacteriales family (*M. formicicum, M. bryantii, M. thermoautotrophicium* and *Methanosphaera stadtmanii*) and 1056 base pairs, from 287 to 1343 base pair in the 16S rDNA, were amplified. The base sequences of 5 clones of gut methanogens and 6 clones of symbiotic bacteria from the protozoa were determined.

Fig. 5 shows alignment of the sequence of *Methanobacterium formicicum,* a symbiont of *Trimyema compressum,* giving an homology ratio of 94%. Thus high homology was obtained for both the gut methanogens and the symbiotic bacteria. The 16S rRNA sequence was closely related to that of *Methanobacterium formicicum.*

SUMMARY

Cellulose digestion in termites is very complex and this complex digestionsupports growth of the termites.The termites themselves, the intestinal protozoa, and the termite intestinal bacteria play an important role in cellulose digestion. We have provided an example of the lower termites playing a role cellulose digestion, but in higher termites, which include the largest number of species in the Termitidae, problems remain to be solved. An important key in obtaining an understanding of the evolution and/or phylogeny of the termite may be found by seeking an understanding of their symbiotic bacteria and the diversity of these bacteria.

ACKNOWLEDGMENTS

This research was supported by a Japan Ministry of Education, Science and Culture Grant-in-Aid for Scientific Research on Priority Areas (#319), Project "Symbiotic Biosphere: An Ecological Interaction Network Promoting the Coexistence of Many Species"

REFERENCES

Cleveland, L. R., 1924, The physiological and symbiotic relationships between the intestinal protozoa of termites and their host, with special reference to *Reticulitermes flavipes* Kollar. Biol. Bull., *46*:178-227

Koga, K., Yamaoka, I., and Murakami, R., 1990, Hydrogenosome in the termite intestinal protozoa *Trichonympha agilis.* Zool. Sci., *7*(6) Suppl.: 1155.

Messer, A. C., and Lee, M. J., 1989, Effect of chemical treatments on methane emission by the hindgut microbiota in the termite *Zootermopsis angusticolis.* Microbial. Ecol., *18*: 275-284.

Shinzato, N., Yamagishi, A., Oshima, T., and Yamaoka, I., 1994, 16S rRNAsequences of methanogenic symbiont in termite. Zool. Sci. *11*(6) Suppl.: 27.

Trager, W., 1934, The cultivation of a cellulose-digesting flagellate, *Trichomonas termopsidis,* and of certain other termite protozoa. Biol. Bull., *66*: 182-190.

Yamaoka, I., 1989, Termite endosymbiosis. In *Insect Endocytobiosis: Morphology, physiology, Genetics, Evolution* ., pp. 77-87. CRC Press, Boca Raton, Florida.

Yamaoka, I., and Murakami, R., 1992, Symbiotic cellulose digestion system in the lower termite: Role of bacteria. Zool. Sci.,*9* (6) Suppl.: 1280.

Yamaoka, I., Murakami, T., and Murakami, R., 1993, Symbiotic cellulose digestion system in the termites: Distribution of methanogenic bacteria. Zool. Sci., *10* (6) Suppl.:162.

Yamaoka, I., and Nagatani, Y., 1975, Cellulose digestion system in the termite, *Reticulitermes speratus* (Kolbe). 1 Producing sites and physiological significance of two kinds of cellulase in the worker. Zool. Mag., *84:* 23-29.

9

THE ROLE OF MICROORGANISMS IN TRI-TROPHIC INTERACTIONS IN SYSTEMS CONSISTING OF PLANTS, HERBIVORES, AND CARNIVORES

Marcel Dicke

Department of Entomology
Wageningen Agricultural University
P.O. Box 8031, 6700 EH Wageningen
The Netherlands

ABSTRACT

Ecosystems comprise populations of individuals at different trophic levels that are connected through many interactions. In studies of insect-plant interactions, macroorganisms have been almost exclusively the focus. Yet, for many of the interactions in multitrophic systems, examples are available showing that microorganisms mediate interactions that were originally thought to occur between macroorganisms. This relates to a variety of aspects, including production of toxins, detoxification of toxins, provisioning of essential nutrients, induction of plant defense mechanism, production of information-conveying chemicals ("infochemicals"), etc. Microorganisms may function as external or internal symbionts, as pathogens, or as mutualists of plants, herbivorous arthropods, or carnivorous arthropods. This review shows that it is essential to investigate interactions between macroorganisms in more detail in order to identify precisely those organisms that interact. In doing so, we obtain a better understanding of the selective forces that operate in ecosystems, because selection in microorganisms results in much faster changes than selection in arthropods and plants, because of the significantly shorter generation time of microorganisms.

INTRODUCTION

The study of insect-plant interactions has long been exclusively concerned with herbivorous insects and their food plants. Such investigations involved, herbivore searching behavior, herbivore feeding behavior, responses of herbivores to secondary plant chemicals, and defensive actions of plants of a chemical and/or physical nature, such as digestibility reducers, toxins, thorns, and cuticle properties. Although interactions between herbivorous

Microbial Diversity in Time and Space, edited by Colwell et al.
Plenum Press, New York, 1996

71

arthropods and their carnivorous enemies were widely studied, it was only since the papers by Bergman and Tingey (1979) and Price et al. (1980) that it was realized that plants could decisively affect such interactions. As a consequence, insect-plant interactions are now more frequently studied in a multitrophic context, mostly for systems consisting of plants, herbivorous arthropods, and carnivorous arthropods. It has been realized that plant characteristics can affect carnivore attack of herbivores, as well as herbivore defense against carnivores.

However, despite this important progress, multitrophic interactions have predominantly been concerned with macroorganisms. Yet, evidence for a role of microorganisms was available, but scattered throughout the literature. During the past decade, several publications have focused on the role of microorganisms in insect-plant interactions and, thus, have brought the subject into the spotlight (Jones, 1984; Price et al., 1986; Martin, 1987; Berenbaum, 1988; Dicke, 1988; Phelan and Stinner, 1991; Barbosa et al., 1991). This was mostly done in the context of plants and herbivorous insects. In this review, are presented examples of the involvement of microorganisms in insect-plant interactions in a tritrophic context, i.e. for interactions of plants, herbivores, and carnivores. This is not an exhaustive review, but rather one that shows that microorganisms may mediate a wide range of interactions between plants, herbivorous and carnivorous arthropods, thereby stimulating entomologists and microbiologists to collaborate in developing this field of research to provide a better understanding of multitrophic interactions.

TRI-TROPHIC INTERACTIONS: PLANTS, HERBIVOROUS ARTHROPODS AND CARNIVOROUS ARTHROPODS

Individuals of consecutive trophic levels are involved in an attack-defense interaction, such as between plants and herbivores or between herbivores and carnivores (Figs. 1a and 1b). Apart from interactions between subsequent trophic levels, each of these interactions may also affect other trophic levels or individuals at one trophic level can affect interactions between members of two other trophic levels. For instance, plant traits may enhance carnivore attack by provision of shelter or alternative food for periods of prey scarcity, or by facilitating carnivore foraging behavior. However, plant characteristics may also interfere with carnivore attack, because of morphological characteristics that hamper carnivore foraging behavior (see Dicke, 1997 for review). On the other hand, plants may affect herbivore defense against their enemies. Plants may enhance herbivore defense, for instance, if the herbivore sequesters its toxins. Alternatively, plants may make herbivores more susceptible to their enemies by prolongation of herbivore development via digestibility reducers or poor nutrient quality (see Price et al., 1980 and Dicke, 1997, for reviews) (Figure 1c).

Microorganisms may be symbionts, mutualists, or pathogens of plants, herbivorous arthropods, or carnivorous arthropods. The term symbiont is used to refer to physiological integration of the microorganism with the plant or arthoropod, without any reference to cost or benefit to either partner. Mutualism refers to an interaction between individuals of the two species that increases the fitness of both, whereas parasitism is used to denote an interaction where the parasite increases its fitness while decreasing the fitness of its host. The sections of this paper are organized by trophic level on which microorganisms function. Not in all cases can we assign a microorganism to one of the main trophic levels, since they can sometimes be regarded as functioning on an intermediate trophic level, thereby causing other trophic levels to be reordered. Therefore, the order of trophic

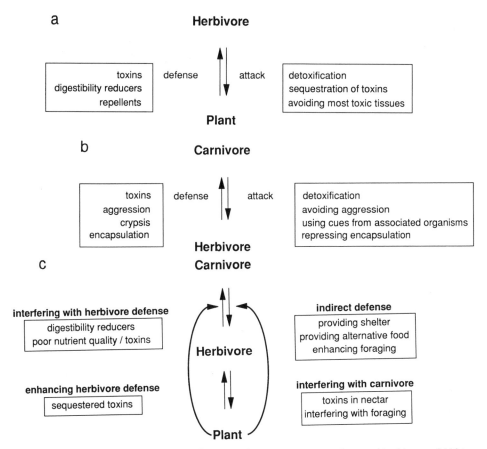

Figure 1. Attack-defense interactions in (a) bi-trophic interactions between plants and herbivores, (b) bi-tro-phic interactions between herbivorous and carnivorous arthropods and (c) tri-trophic interactions, with emphasis on the effects of plants on attack and defense in herbivore-carnivore interactions.

levels is restricted so that plants are always on the first level, herbivores on the second, and carnivores on the third.

MICROORGANISMS AT THE FIRST TROPHIC LEVEL

Microorganisms that are a true component of the first trophic level are saprophytes, such as fungi and mushrooms that serve as food for flies and beetles. In other situations, the position of the microorganism can be considered more intermediate, i.e., in between the plant and the herbivore. Examples are plant-infecting yeasts, bacteria and fungi that create a feeding site for insects, such as flies or beetles (Dicke, 1988; Drew and Lloyd, 1991).

A first step in the location of food by insects is long-distance orientation. Insects that feed on saprophytic microorganisms or on microbe-infected plants are well-known to respond to volatile infochemicals [chemicals that convey information, *sensu* Dicke and Sabelis (1988)] during this foraging phase (Dicke, 1988). For instance, *Drosophila melanogaster* flies are strongly attracted to the odor of fermenting fruits, especially to volatiles produced by microorganisms infecting the fruit. Among the attractive volatiles are ethanol, acetic acid, lactic acid, *n*-butyraldehyde, ethyl acetate, diacetyl, acetoin, acetaldehyde, and indole (Hutner et al., 1937; Fuyama, 1976). Once at the food site, the flies feed and oviposit. The larvae of *Drosophila* flies are subject to parasitization by parasitic wasps (parasitoids). The wasps lay their eggs in the fly larvae and the emerging wasp larvae develop within the fly larvae and, finally, adult wasps emerge from the fly pupae. During foraging for hosts, these parasitic wasps rely to a large extent on microbial volatiles also (Vet, 1985). For instance, the parasitic wasp *Leptopilina heterotoma*, that can parasitize *D. melanogaster*, is attracted to ethanol, acetaldehyde, and ethyl acetate (Dicke et al., 1984). It is significant to note that although these fermentation products attract the insects, the fly and wasp need to detoxify the products in order to survive exposure to these chemicals once they arrive and stay at the fermented fruit. *Drosophila* flies tolerate relatively high concentrations of fermentation products, such as ethanol, acetic acid, and acetaldehyde, and even use these substances as a resource (Moxon et al., 1982; Parsons, 1981, 1982). Their parasitoids also have an increased, though lower, tolerance to ethanol, which is especially true for the females (Bouletreau and David, 1981). This higher tolerance is functional because females spend a significant amount of time searching on the fermenting substrate while males never return to this environment after eclosion as adults. Thus, the volatiles produced by the microbial activity on food of *Drosophila* spp. affect the behavior and physiology of the microbivore, as well as its enemies.

The attraction of microbivores has also been observed for insects that feed on saprophages, such as mushrooms (Grove and Blight, 1983; Pfeil and Muma, 1993). To my knowledge the only information about the role of microbial volatiles in the attraction of natural enemies of the insects that are attracted to saprophage odors relates to parasitic wasps that attack fungivorous *Drosophila* (Tieleman, 1988).

In order to obtain their food, natural enemies of microbivores may even employ microbial odors to attract the microbivores. The social spider, *Mallos gregalis*, attracts its prey with a scented bait that consists of discarded carcasses of partially eaten muscoid flies on which yeasts are growing (Tietjen et al., 1987). Feeding activities of the spider condition the prey carcass, such that yeasts, rather than bacteria, flourish. The attracted prey consists mostly of flies, among which are those that feed on a wide range of decaying substances.

Microbivores or insects that feed on microbe-infected plant material search for diseased plants. Their development is significantly enhanced by the microbes or their products and, in many cases they can only survive when the microbes are present (Dicke, 1988). In contrast, true herbivores are often negatively affected by microbial infestation of their food plants. Yet, in most cases, microbial infection is not beneficial to the plant, since it reduces its fitness. However, the situation appears to be quite different with clavicipitaceous fungal endophytes of grasses. These microbes produce alkaloids that defend their grass host against herbivory, while increasing the host's fitness (Dahlman et al., 1991; Clay, 1991; Carroll, 1991). The increase in fitness, being obvious in the presence of herbivores, is also evident in the absence of herbivores. Infected plants may produce more biomass than uninfected plants. The mechanisms underlying this effect are not known but may include alterations of phytohormone levels, stomatal conductance etc. (Clay, 1991). The effects of endophytic fungi on herbivores comprise decreased survival and growth and increased developmental time (Clay, 1991; Carroll, 1991). The latter results in the herbivores being exposed longer to their predators, parasitoids, and pathogens. Thus, the endophyte is considered mutualistic with the plant it infects and this mutualism affects not only herbivorous insects, but also their natural enemies. The interaction between plants and endophytic fungi clearly shows how microbiological information can provide a vital contribution to the understanding of interactions in plant-herbivore-carnivore systems.

In the previous paragraphs, the microbivores feeding on plant-related microorganisms were all at the second trophic level. Yet, facultative microbivores can also be found at the third trophic level. Many carnivorous arthropods can feed on a variety of food sources. When their preferred prey is absent or scarce, the carnivores may rely on food sources, such as extrafloral nectar, pollen, or microorganisms. For instance, several species of predatory mites that prefer herbivorous mites as food accept plant-contaminating fungi as food when the prey mites are not available (Overmeer, 1985; Bakker and Klein, 1992). Thus, plant-associated microorganisms may affect the abundance of carnivorous arthropods on a plant and, thereby, influence herbivorous mite populations.

MICROORGANISMS AT THE SECOND TROPHIC LEVEL

Herbivore Symbionts and Mutualists

Plants are well-known for the variety of secondary chemicals that can be found in them, many of which are toxic to insects or result in lower digestibility. The previous section has shown that microorganisms may be responsible for these characteristics, as in the case of clavicipitaceous fungal endophytes. In addition, plants often do not contain a complete diet for insects, due to the absence of essential nutrients. Some herbivorous arthropods cope with these suboptimal qualities of their food by introducing microorganisms that detoxify secondary plant chemicals or predigest the plant tissue (Buchner, 1953; Martin, 1987; Berenbaum, 1988). For instance, bark beetles or woodwasps carry fungi parasitic on trees, in special morphological structures, called mycangia (Leufven, 1991). After introduction into the tree, the fungi help the insects in subduing the tree and in conditioning the plant material such that the herbivore can successfully utilize it (Leufven, 1991). Thus, in contrast to the situation for microbivores (previous section), microorganisms can here be regarded as external variants of herbivore symbionts. Enemies of these herbivores are known to use chemical cues from the microorganisms to locate the herbivores (Madden, 1968; Spradbery, 1970; Wood, 1982). In more extreme cases, often occurring in leaf hoppers and aphids, the herbivorous insect is a vector of a plant pathogenic microorganism, usually a virus. Also in these situations, the microbe may increase the nutritional value of the plant for the herbivorous insects (Carter, 1973). For instance, the introduction of yellow mosaic virus into zucchini plants by the aphid *Aphis gossypii* affects the nutrient status of the host plant, such that the intrinsic rate of population increase of the aphids is enhanced (Blua et al., 1994).

Other herbivorous insects contain intracellular symbiotic microorganisms that provide the insect with essential nutrients (Buchner, 1953; Houk, 1987; Berenbaum, 1988). Moreover, the catabolic abilities of many symbiotic microorganisms of herbivorous insects indicate that they are also able to detoxify secondary plant chemicals (Jones, 1984). This indeed has been demonstrated in several cases (Dowd, 1991). For example, the cigarette beetle *Lasioderma serricorne* harbors symbiotic microorganisms that can detoxify a range of secondary plant chemicals, such as tannic acid, 1-naphthyl acetate, *trans*-cinnamic acid, gallic acid, morin, quercetin, trihydroflavone, and umbelliferone (Dowd, 1991). Herbivores that harbor such detoxifying microorganisms have a different effect on their natural enemies than herbivores that sequester secondary plant chemicals. Thus, these detoxifying symbionts also affect the defense of the herbivores to their carnivorous enemies.

Extreme cases of herbivore-microorganism mutualism are found in the association of ants and termites with fungi (Cherrett et al., 1989; Wood and Thomas 1989). These herbivores have special fungus gardens in which the fungus is provided with plant food where predigestion takes place. The herbivores provide the fungi with protected environments and they suppress the germination and growth of other fungi while the fungi predigest the food for the herbivores.

In the process of detoxifying secondary plant chemicals, symbiotic microorganisms also produce certain waste products. For instance, in detoxifying sulfur-containing plant chemicals, symbionts of caterpillars of the leek moth, *Acrolepiopsis assectella*, produce sulfur-containing volatiles that are emitted from the caterpillar feces. These odors are used by parasitoids of the caterpillars to locate their victims. These volatiles attract the parasitoids to feeding sites of the caterpillars and, thereby, increase the risk of parasitization (Thibout et al., 1993). Among the bacteria found in the caterpillars' feces are several *Bacillus* species, an *Enterobacter* species, and a *Staphylococcus* species. The advantage of detoxification of leek secondary chemicals by bacteria apparently outweighs the disadvantage of attracting parasitic wasps. Bark beetle feces also appears to be a source of microbial infochemicals. Bark beetles are well-known for microbial symbionts involved in detoxifying plant terpenoids (Leufven, 1991). Feces of bark beetles emit chemicals that attract conspecifics to an infested tree. Below a certain density this attraction of conspecifics is advantageous to the resident beetles because communal attack increases the chance of subdueing the tree. Some of the chemicals that attract conspecific bark beetles are suspected to be produced by microorganisms in the gut of the beetle (Brand et al., 1975). The fecal odors not only attract conspecifics but also predators and parasitic wasps of the beetles (Wood, 1982). Fecal odors are known to play a role in the foraging behavior of many parasitoid species, especially in short-distance searching (Weseloh, 1981). However, the origin of the infochemicals has not been fully investigated. The work of Thibout et al. (1993) should stimulate studies on this subject. An important question is whether the influence of microbial volatiles in feces is a general phenomenon or whether it is especially related to herbivores that feed on plants that are defended by specific groups of secondary plant chemicals, such as sulfur-containing chemicals in *Allium* species, glucosinolates in crucifers, or terpenoids in pine trees.

Plant Pathogens

Plants are known to respond to herbivory and to pathogen infestation by induced production of chemicals that interfere with the attacker's development or reproduction (Agrios, 1988; Tallamy and Raupp, 1991). This effect is often not very specific. Infection with one pathogen has effects on conspecific as well as heterospecific pathogens and the same holds for herbivores. Moreover, often cross-effects have been reported for pathogen infection on subsequent herbivore infection and *vice-versa*. For example, the fungal pathogen *Verticillium dahliae* was less likely to cause symptoms of verticillium wilt on cotton seedlings that had been previously exposed to spider mites than on unexposed cotton seedlings. Conversely, populations of the spider mite *Tetranychus urticae* grew less rapidly on seedlings that had been inoculated with *V. dahliae* than on uninoculated controls (Karban et al., 1987). Some data even suggest that the response of plants to herbivory may actually be a response to microorganisms that contaminate the herbivore mouthparts or secretions (Hartley and Lawton, 1991).

In addition to non-volatile chemicals induced by infection of plants, herbivory and pathogen attack can result in the induced production of plant volatiles. These volatiles may also have a negative effect on herbivore or pathogen performance. For instance, plant volatiles that result from the lipoxygenase pathway and are emitted in response to either herbivore damage (Visser 1986) or pathogen infection (Croft et al., 1993) comprise six-carbon aldehydes that reduce aphid reproduction (Hildebrand et al., 1993), as well as bacterial proliferation (Zeringue and McCormick, 1989; Deng et al. 1993). The effects of herbivore-induced plant volatiles on a range of organisms of different trophic levels has been under intensive study during the past 10 years. Herbivore-induced plant chemicals often comprise *de novo* synthesized terpenoids (Dicke, 1994). These chemicals attract carnivorous enemies of the herbivores, they may deter herbivores and they may even affect neighboring plants

(Bruin et al., 1992; Dicke et al., 1990, 1993; Turlings et al., 1993). It is well-known that microbial elicitors trigger plant defensive responses and the chemical nature of many of these elicitors has been elucidated (de Wit, 1992). In recent years it has also been shown that herbivore secretions contain elicitors that trigger plant defensive responses (Lin et al., 1990; Turlings et al., 1990; Mattiacci et al., 1994). Recently, the first herbivore-secreted elicitor has been identified as the enzyme β-glucosidase (Mattiacci et al., 1995) but its exact origin remains unknown. It has been suggested that herbivore-secreted elicitors may be of microbial origin, because of the similarity of many plant responses to herbivores and microbes, where microbe-induced responses may also affect herbivore performance and vice versa (Schultz, 1993). In this respect, it is interesting to know that tephritid flies introduce bacteria into a plant while regurgitating (Drew and Lloyd, 1991). Further research is needed to explore whether microorganisms are involved in plant responses to herbivory. It seems that if microorganisms are involved, they should produce specific elicitors in this case, since the induced plant volatiles may differ when different herbivore species infest plants of the same species (Turlings et al., 1993; Dicke, 1994). Such specific elicitors can certainly be present, since the production of plant volatiles in response to pathogen attack can depend on pathogen identity (Croft et al., 1993).

Plant pathogens may be food for arthropods. This relates to interactions discussed in the section 'Microorganisms at the first trophic level'.

MICROORGANISMS AT THE THIRD TROPHIC LEVEL

Carnivore Symbionts and Mutualists

Carnivores may encounter secondary plant chemicals that have been sequestered by their herbivorous victim. Just as herbivores may harbor symbionts that detoxify the plant chemicals, carnivorous arthropods might also employ symbionts for this type of activity. However, no example of this has been published. Yet, carnivores do harbor symbiotic microorganisms that help them to overcome herbivore defenses. Parasitic wasps that deposit their eggs into a herbivorous host may lose their offspring as a result of encapsulation of wasp eggs by the herbivore. Encapsulation is a cellular defense whereby host haemocytes form an enveloping multicellular capsule around parasitoid eggs or larvae. Parasitoids may suppress encapsulation through the aid of viruses that are injected into the host during oviposition (Edson et al., 1981; Stolz and Vinson, 1979; Fleming, 1992). Elegant studies have shown that these viruses are present in calyx tissue along the oviduct of the parasitoid. The viruses are added to the eggs when they pass through the oviduct. The effect of the viruses is not specific for a certain parasitic wasp species. Introduction of the virus of one wasp species will also protect eggs of other wasp species (Vinson and Stoltz, 1986). Therefore, the virus appears to affect the production of the capsule.

Another remarkable phenomenon of carnivore symbionts has been discovered recently. Parasitic wasps may harbor symbionts that make the wasps reproduce asexually: unmated, infected females produce female offspring exclusively, whereas uninfected females need mating in order to produce female offspring and they always produce sons as well (Stouthamer et al., 1993; Stouthamer and Werren, 1993). Thus, the symbionts affect the rate of population increase of the carnivores and consequently influence herbivore populations. The symbionts, *Wolbachia* spp., are exclusively transferred through the cytoplasm of the eggs from mother to daughter; males constitute a dead-end street to the microbes (Stouthamer et al., 1993). Thus, it is obviously to the benefit of the symbiont if all offspring of infected mothers are daughters. To the carnivore, all-female offspring is obviously also advantageous, unless there is a serious fitness cost to being infected by *Wolbachia*. This appears to be the case at unlimited host availability: uninfected, sexually reproducing wasps in the genus *Trichogramma* produce more offspring, in several cases even more female offspring than infected asexually reproducing conspecifics. However, in host-limited situations the infected wasps do much better than uninfected wasps: they produced more daughters than uninfected wasps (Stouthamer and Luck, 1993).

Carnivores may be exposed to plant secondary chemicals in two ways. Either indirectly through their herbivorous prey that may have sequestered the chemicals, or directly by feeding on plant tissues or secretions such as extrafloral nectar (Dicke, 1997). This may be a threat to the carnivore symbiont and, thus, it may affect food choice. Future investigations should take this into account.

Herbivore Pathogens

Herbivore pathogens may result in a range of malfunctions in the herbivorous host, ranging from behavioral deformations such as non-responsiveness to sex-pheromones (Sweeney and McLean, 1987) to developmental retardation and death (Tanada and Kaya, 1993). These effects also influence interactions between the herbivore and its foodplant and between the herbivore and its carnivorous arthropod enemies. Especially with regard to the latter, interesting phenomena are known. The effects of diseased herbivores on their accept-ability to predators or parasitic wasps can vary from neutral to negative (Vinson and Iwantsch, 1980; Brooks, 1993) and they can even negatively affect carnivore reproduction (Huger, 1984). Moreover, an interesting phenomenon is that the pathogen may alter herbi-vore behavior to its own benefit. For instance, the infected herbivore can be induced to move to exposed sites, which increases predation chances and, consequently, dispersal of the pathogens (Holmes and Bethel, 1972; Horton and Moore, 1993). However, if the host is under control of its own behavior, there may be circumstances that favor the host to commit suicide so as to avoid infection of closely related individuals (McAllister and Roitberg, 1987).

Herbivores may defend themselves against pathogens by sequestering secondary plant chemicals. A well known example is that of tobacco horn worm larvae (*Manduca sexta*) that gain protection from feeding on nicotine-containing plants. With increasing nicotine content of their diet, they suffer less mortality from the bacterium *Bacillus thuringiensis*. In contrast, herbivores that are sensitive to nicotine, such as caterpillars of the generalist *Trichoplusia ni* suffer mortality from feeding on nicotine-containing diets and do not gain protection against pathogens (Krischik et al., 1988; Krischik, 1991). Similar examples have also been recorded for the effect of digestibility reducers such as tannins, on the susceptibility of caterpillars to viruses or *B. thuringiensis* endotoxin (Navon et al., 1993; Schultz and Keating, 1991; Hunter and Schultz, 1993).

MICROORGANISMS AT THE FOURTH TROPHIC LEVEL

The knowledge of pathogens of carnivorous arthropods is limited, compared to that of herbivore pathogens, but several cases have been reported (Hess and Hoy, 1982; Hamm et al., 1988). This relatively limited information is probably due to the fact that pathogens of herbivores have been studied with the aim of employing them in biological control programs (Smits, 1994). However, basically similar effects can be expected, as known for the influence of pathogens on herbivores, to be discovered when research is carried out on carnivore-pathogen interactions. Thus, effects of carnivore pathogens may affect carnivore-herbivore interactions. For example, the parasitic wasp *Microplitis croceipes* may suffer from a viral infection that affects its flight behavior, with consequently impaired success of finding herbivorous hosts (Hamm et al., 1988).

CONSEQUENCES FOR UNDERSTANDING OF THE FUNCTIONING OF MULTI-TROPHIC SYSTEMS

The study of plant-arthropod interactions has yielded many ideas about evolution of plant defense (Ehrlich and Raven, 1964; Feeny, 1976; Coley et al., 1985; Rhoades, 1985; Price, 1991), herbivore attack, and degree of specialization of herbivores (Bernays and Graham, 1988; Rausher, 1988; Thompson 1988) and carnivore foraging strategies (Price, 1991; Vet and Dicke, 1992). Selective pressures on the macroorganisms have been considered exclusively. However, it now appears from the literature that interactions between macroorganisms may be decisively influenced by microorganisms that are their symbionts, mutualists, or pathogens. In fact, the interactions in tri-trophic systems consisting of plants, and herbivorous and carnivorous arthropods are much more complex than depicted in Figure 1c: see Figure 2 for a summary of all interactions that have been reviewed in this paper. This has tremendous consequences with regard to evolutionary forces operating in multitrophic interactions. For instance, if herbivores adapt to plant chemical defenses through detoxifying microorganisms, then adaptation may occur at a much higher rate, due to the considerably shorter generation time of microorganisms (Jones, 1984; Price et al., 1986; Berenbaum 1988). Furthermore, this raises questions such as why do some insect species adapt to a certain plant species

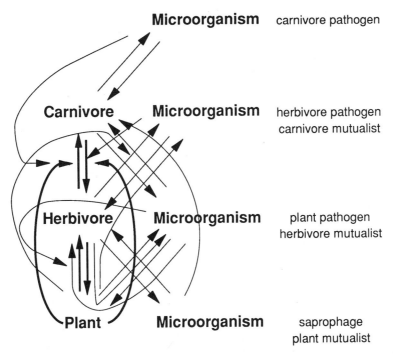

Figure 2. Interactions between microorganisms and macroorganisms in tri-trophic systems consisting of plants, herbivorous arthropods, and carnivorous arthropods. Each of the arrows relates to a phenomenon treated in this review.

while others do not? Is this determined by their symbionts and if so, why do their symbionts differ?

This review has provided examples of microbial mediation of interactions in multitrophic interactions. The examples are derived from a range of insect orders. Not all examples have been recorded for the same insect order. For instance, detoxification of secondary plant chemicals by intracellular symbionts has been predominantly reported for the orders Orthoptera, Homoptera, and Coleoptera, while information on foliage-chewing herbivorous insects is limited (Dowd, 1991). To evaluate better the impact of microorganisms, comparative studies are needed within and between insect orders. Comparisons between insect orders should be carried out to identify whether microbial influences are more prominent in some orders than others and why this is so. Within order comparisons should be made between closely related species that differ in certain ecological characteristics, such as the degree of specialization on a certain food source, or the pressure from enemies. Such studies, carried out collaboratively by botanists, entomologists, and microbiologists, will make an enormous contribution to our understanding of multitrophic interactions.

ACKNOWLEDGMENTS

I thank Prof. Dr. U. Simidu for the invitation to write this chapter and Dr. Richard Stouthamer for comments on the manuscript.

REFERENCES

Agrios, G.N. 1988. How plants defend themselves against pathogens. In: G.N. Agrios, Plant Pathology, Third Edition, pp. 97-115. Academic Press, San Diego.

Bakker, F.M. and Klein, M.E. 1992. Transtrophic interactions in cassava. Exp. Appl. Acar. 14: 293-311.

Barbosa, P., Krischik, V.A. and Jones, C.G. (eds.), 1991. Microbial Mediation of Plant-Herbivore Interactions. John Wiley and Sons, New York, 530 p.

Brand, J.M., Bracke, J.W., Markovetz, A.J., Wood, D.L. and Browne, L.E. 1975. Production of verbenol pheromone by a bacterium isolated from bark beetles. Nature 254: 136-137.

Berenbaum, M.R. 1988. Allelochemicals in insect-microbe-plant interactions; agents provocateurs in the coevolutionary arms race. In: Barbosa, P. and Letourneau D.K. (eds.) Novel Aspects of Insect-Plant Interactions. pp. 97-123. Wiley and Sons, New York.

Bergman, J.M. and Tingey, W.M. 1979. Aspects of interaction between plant genotypes and biological control. Bull. Ent. Soc. Amer. 25,275-279.

Bernays, E. and Graham, M. 1988. On the evolution of host specificity in phytophagous arthropods. Ecology 69: 886-892.

Blua, M.J., Perring, T.M. and Madore, M.A. 1994. Plant virus-induced changes in aphid population development and temporal fluctuations in plant nutrients. J. Chem. Ecol. 20: 691-707.

Bouletreau, M. and David, J.R. 1981. Sexually dimorphic response to host habitat toxicity in Drosophila parasitic wasps. Evolution 35: 395-399.

Brooks, W.M. 1993. Host-parasitoid-pathogen interactions. In: N.E. Beckage, S.N. Thompson and B.A. Federici (eds.) Parasites and Pathogens of Insects. Vol. 2: 231-272. Acad Press, San Diego.

Bruin, J., Dicke, M. and Sabelis, M.W. 1992. Plants are better protected against spider-mites after exposure to volatiles from infested conspecifics. Experientia 48: 525-529.

Buchner, P. 1953. Endosymbiose der Tiere mit pflanzlichen Mikroorganismen. Birkhaeuser, Basel. 771 pp.

Carroll, G.C. 1991. Fungal associates of woody plants as insect antagonists in leaves and stems. In: P. Barbosa, V.A. Krischik and C.G. Jones (eds.), Microbial mediation of plant-herbivore interactions, pp. 253-271, John Wiley and Sons, New York.

Carter, W. 1973. Insects in relation to plant disease. Wiley Interscience, New York.

Cherrett, J.M., Powell, R.J. and Stradling, D.J. 1989. The mutualism between leaf-cutting ants and their fungus. In: N. Wilding, N.M. Collings, P.M. Hammond and J.F. Webber (eds.) Insect-Fungus Interactions. pp. 93-120. Acad. Press.

Clay, K. 1991. Fungal endophytes, grasses, and herbivores. In: P. Barbosa, V.A. Krischik and C.G. Jones (eds.), Microbial mediation of plant-herbivore interactions, pp. 199-226, John Wiley and Sons, New York.

Coley, P.D.; Bryant, J.P. and Chapin, F.S. 1985. Resource availability and plant antiherbivore defense. Science 230: 895-899.

Croft, K.P., Juttner, F. and Slusarenko, A.J. 1993. Volatile products of the lipoxygenase pathway evolved from *Phaseolus vulgaris* (L.) leaves inoculated with *Pseudomonas syringae pv phaseolicola*. Plant Physiol. 101: 13-24.

Dahlman, D.L., Eichenseer, H. and Siegel, M.R. 1991. Chemical perspectives on endophyte-grass interactions and their implications to insect herbivory. In: P. Barbosa, V.A. Krischik and C.G. Jones (eds.), Microbial mediation of plant-herbivore interactions, pp. 227-252, John Wiley and Sons, New York.

Deng, W., Hamilton-Kemp, T.R., Nilesen, M.T., Andersen, R.A., Collins, G.B. and Hildebrand, D.F. 1993. Effects of six-carbon aldehydes and alcohols on bacterial proliferation. J. Agric. Food Chem. 41: 506-510.

Dicke, M. 1988. Microbial allelochemicals affecting the behavior of insects, mites, nematodes and protozoa in different trophic levels. In: Barbosa, P. and Letourneau, D. (eds.) Novel Aspects of Insect-Plant Interactions. John Wiley and Sons, New York. pp. 125-163.

Dicke, M. 1994. Local and systemic production of volatile herbivore-induced terpenoids: Their role in plant-carnivore mutualism. J. Plant Physiol. 143: 465-472.

Dicke, M. 1997. Direct and indirect effects of plants on beneficial organisms. In: J.R. Ruberson (ed.) Handbook of Pest Management. Marcel Dekker, Inc, New York. (in press).

Dicke, M. and Sabelis M.W. 1988. Infochemical terminology: based on cost-benefit analysis rather than origin of compounds? Funct. Ecol. 2: 131-139.

Dicke, M., Bruin, J. and Sabelis, M.W. 1993. Herbivore-induced plant volatiles mediate plant-carnivore, plant-herbivore and plant-plant interactions: Talking plants revisited. In: J.C. Schultz and I. Raskin (eds.) Plant signals in interactions with other organisms. Current Topics in Plant Physiology, An American Society of Plant Physiologists Series Vol. 11: 182-196

Dicke, M., Lenteren, J.C. van, Boskamp, G.J.F. and Dongen-van Leeuwen, E. van, 1984. Chemical stimuli in host-habitat location by *Leptopilina heterotoma* (Thomson) (Hymenoptera: Eucoilidae), a parasite of *Drosophila*. J. Chem. Ecol. 10: 695-712.

Dicke, M., Sabelis, M.W., Takabayashi, J., Bruin, J. and Posthumus, M.A. 1990. Plant strategies of manipulating predator-prey interactions through allelochemicals: prospects for application in pest control. J. Chem. Ecol. 16: 3091-3118.

Dowd, P.F. 1991. Symbiont-mediated detoxification in insect herbivores. In: P. Barbosa, V.A. Krischik and C.G. Jones (eds.), Microbial mediation of plant-herbivore interactions, pp. 411-440, John Wiley and Sons, New York.

Drew, R.A.I. and Lloyd, A.C. 1991. Bacteria in the life cycle of tephritid fruit flies. In: P. Barbosa, V.A. Krischik and C.G. Jones (eds.), Microbial mediation of plant-herbivore interactions, pp. 441-465, John Wiley and Sons, New York.

Edson, K.M., Vinson, S.B., Stoltz, D.B. and Summers, M.D. 1981. Virus in a parasitoid wasp: suppression of the cellular immune response in the parasitoid's host. Science 211: 582-583.

Ehrlich, P. and Raven, P.H. 1964. Butterflies and plants: A study in coevolution. Evolution 18: 586-608.

Feeny, P. 1976. Plant apparency and chemical defense. Rec. Adv. Phytochem 10: 1-40.

Fleming, J.-A. G.W. 1992. Polydnaviruses: Mutualists and Pathogens. Annu.Rev. Entomol. 37: 401-425.

Fuyama, Y. 1976. Behavior genetics of olfactory responses in *Drosophila*. I. Olfactometry and strain differences in *Drosophila melanogaster*. Behav. Genet. 6: 407-420.

Grove, J.F. and Blight, M.M. 1983. The oviposition attractant for the mushroom phorid *Megaselia halterata*: the identification of volatiles present in mushroom house air. J. Sci. Food Agric. 34: 181-185.

Hamm, J.J., Styer, E.L. and Lewis, W.J. 1988. A baculovirus pathogenic to the parasitoid *Microplitis croceipes* (Hymenoptera: Braconidae). J. Invert. Pathol. 52: 189-191.

Hartley, S.E. and Lawton, J.H. 1991. Biochemical aspects and significance of the rapidly induced accumulation of phenolics in birch foliage. In: D.W. Tallamy and M.J. Raupp (eds.), Phytochemical Induction by Herbivores, pp. 105-132, John Wiley and Sons, New York.

Hess, R.T. and Hoy, M.A. 1982. Microorganisms associated with the spider mite predator *Metaseiulus* (= *Typhlodromus*) *occidentalis*: Electron microscope observations. J. Invert. Pathol. 40: 98-106.

Hildebrand, D.F., Brown, G.C., Jackson, D.M.and Hamilton-Kemp, T.R. 1993. Effects of some leaf-emitted volatile compounds on aphid population increase. J. Chem. Ecol. 19: 1875-1887.

Holmes, J.C. and Bethel, W.M. 1972. Modification of intermediate host behavior by parasites. J. Linn. Soc. (Zool.) (Suppl. 1) 51: 123-149.

Horton, D.R. and Moore, J. 1993. Behavioral effects of parasites and pathogens in insect hosts. In: N.E. Beckage, S.N. Thompson and B.A. Federici (eds.) Parasites and Pathogens of Insects. Vol. 1: 107-124. Acad. Press, San Diego.

Houk, E.J. 1987. Symbionts. In: A.K. Minks and P. Harrewijn (eds.) Aphids. Their Biology, Natural Enemies and Control. World Crop Pests 2A: 123-129. Elsevier, Amsterdam.

Huger, A.M. 1984. Susceptibility of the egg parasitoid *Trichogramma evanescens* to the microsporidium *Nosema pyrausta* and its impact on fecundity. J. Inv. Pathol. 44: 228-229.

Hunter, M.D. and Schultz, J.C. 1993. Induced plant defenses breached? Phytochemical induction protects an herbivore from disease. Oecologia 94: 195-203.

Hutner, S.H., Kaplan, H.M. and Enzmann, E.V. 1937. Chemicals attracting *Drosophila*. American naturalist 71: 575-581.

Jones, C.G. 1984. Microorganisms as mediators of plant resource exploitation by insect herbivores. In: P.W.Price, C.N. Slobodchikoff and W.S. Gaud (eds.) A new ecology. Novel approaches to interactive systems, pp. 53-99. John Wiley and Sons, New York.

Karban, R., Adamchak, R. and Schnathorst, W.C. 1987. Induced resistance and interspecific competition between spider mites and a vascular wilt fungus. Science 235: 678-680.

Krischik, V.A. 1991. Specific or generalized plant defense: reciprocal interactions between herbivores and pathogens. In: P. Barbosa, V.A. Krischik and C.G. Jones (eds.), Microbial mediation of plant-herbivore interactions, pp. 309-340, John Wiley and Sons, New York.

Krischik, V.A., Barbosa, P. and Reichelderfer, C.F. 1988. Three trophic level interactions: allelochemicals, *Manduca sexta* (L.), and *Bacillus thuringiensis var. kurstaki* Berliner. Environ. Entomol. 17: 476-482.

Leufven, A. 1991. Role of microorganisms in spruce bark beetle-conifer interactions. In: P. Barbosa, V.A. Krischik and C.G. Jones (eds.), Microbial mediation of plant-herbivore interactions, pp. 467-483, John Wiley and Sons, New York.

Lin, H., Kogan, M. and Fischer, D. 1990. Induced resistance in soybean to the Mexican bean beetle (Coleptera: Coccinellidae): comparison of inducing factors. Environ. Entomol. 19: 1852-1857.

Madden, J.L. 1968. Behavioural responses of parasites to the symbiotic fungus associated with *Sirex noctilio* F. Nature 218: 189-190.

Martin, M.M. 1987. Invertebrate-Microbial Interactions: Ingeste Fungal Enzymes in Arthoropod Biology. Cornell University Press, Ithaca NY, 148 pp.

Mattiacci, L., Dicke, M. and Posthumus, M.A. 1994. Induction of parasitoid attracting synomone in Brussels sprouts plants by feeding of *Pieris brassicae* larvae: role of mechanical damage and herbivore elicitor. J. Chem. Ecol. 20: 2229-2247.

Mattiacci, L., Dicke, M. and Posthumus, M.A. 1995. β-glucosidase: elicitor of herbivore-induced plant odors that attract host-searching parasitic wasps. Proc. Natl. Acad. Sci. USA 92: 2036-2040.

McAllister, M.K. and Roitberg, B.D. 1987. Adaptive suicidal behaviour in pea aphids. Nature 328: 797-799.

Moxon, L.N., Holmes, R.S. and Parsons, P.A. 1982. Comparative studies of aldehyde oxidase, alcohol dehydrogenase and aldehyde resource utilization among Australian *Drosophila* species. Comp. Biochem. Physiol. 71: 387-395.

Navon, A., Hare, J.D. and Federici, B.A. 1993. Interactions among *Heliothis virescens* larvae, cotton condensed tannin and the CryIA(c) delta-endotoxin of *Bacillus thuringiensis*. J. Chem. Ecol. 19: 2485-2499.

Overmeer, W.P.J. 1985. Alternative prey and other food resources. In: W. Helle and M.W. Sabelis (eds.) Spider Mites. Their Biology, Natural Enemies and Control. World Crop Pests 1B: 131-139. Elsevier, Amsterdam.

Parsons, P.A. 1981. Longevity of cosmopolitan and native Australian *Drosophila* in ethanol atmospheres. Aust. J. Zool. 29: 33-39.

Parsons, P.A. 1982. Acetic acid vapour as a resource and stress in *Drosophila*. Aust. J. Zool. 30: 427-433.

Phelan, P.L. and Stinner, B.R. 1991. Microbial mediation of plant-herbivore ecology. In: G.A. Rosenthal and M.R. Berenbaum (eds.) Herbivores: their interaction with secondary plant metabolites. Second edition. Vol 2: 279- 315. Acad. Press, New York.

Pfeil, R.M. and Muma, R.O. 1993. Bioassay for evaluating attraction of the phorid fly, *Megaselius halterata* to compost colonized by the commercial mushroom *Agaricus bisporus* and to 1-octen-3-ol and 3-octanone. Entomol. Exp. Appl. 69: 137-144.

Powell, W., Wilding, N., Brobyn, P.J. and Clark, S.J. 1986. Interference between parasitoids (Hym.: Aphidiidae) and fungi (Entomophthorales) attacking cereal aphids. Entomophaga 31: 293-302.

Price, P.W. 1991. Evolutionary theory of host and parasitoid interactions. Biol. Contr. 1: 83-93.

Price, P.W., Bouton, C.E., Gross, P., McPheron, B.A., Thompson, J.N. and Weis, A.E. 1980. Interactions among three trophic levels: influence of plant on interactions between insect herbivores and natural enemies. Annu. Rev. Ecol. Syst. 11: 41-65.

Price, P.W., Westoby, M., Rice, B., Atsatt, P.R., Fritz, R.S., Thompson, J.N. and Mobley, K. 1986. Parasite mediation in ecological interactions. Annu. Rev. Ecol. Syst. 17: 487-505.

Rausher, M.D. 1988. Is coevolution dead? Ecology 69: 898-901.

Rhoades, D.F. 1985. Offensive-defensive interactions between herbivores and plants: their relevance in herbivore population dynamics and ecological theory. Am. Nat. 125: 205-238.

Schultz, J.C. 1993. Signaling in plant responses to herbivory. In: J.C. Schultz and I. Raskin (eds.) Plant signals in interactions with other organisms. Current Topics in Plant Physiology, An American Society of Plant Physiologists Series Vol. 11: 93-101.

Schultz, J.C. and Keating, S.T. 1991. Host-plant-mediated interactions between te gypsy moth and a baculovirus. In: P. Barbosa, V.A. Krischik and C.G. Jones (eds.), Microbial mediation of plant-herbivore interactions, pp. 489-506, John Wiley and Sons, New York.

Smits, P.H. (ed.) 1994. Microbial Control of Pests. IOBC/WPRS Bull. 17(3): 1-307.

Spradbery, J.P. 1970. Host finding by *Rhyssa persuasoria* (L.), an ichneumonid parasite of siricid woodwasps. Anim. Behav. 18: 103-114

Stoltz, D.B. and Vinson, S.B. 1979. Viruses and parasitism in insects. Adv. Virus Res. 24: 125-171.

Stouthamer, R. and Luck, R.F. 1993. Influence of micro-associated parthenogenesis on the fecundity of *Trichogramma deion* and *T. pretiosum*. Entomol. Exp. Appl. 67: 183-192.

Stouthamer, R. and Werren, J.H. 1993. Microbes associated with parthenogenesis in wasps of the genus *Trichogramma*. J. Inv. Pathol. 61: 6-9.

Stouthamer, R., Breeuwer, J.A.J., Luck, R.F. and Werren, J.H. 1993. Molecular identification of microorganisms associated with parthenogenesis. Nature 361: 66-68.

Sweeney, J.D. and McLean, J.A. 1987. Effect of sublethal infection levels of *Nosema* sp. on the pheromone-mediated behavior of the western spruce budworm, *Choristoneura occidentalis* Freeman (Lepidoptera: Tortricidae). Can. Ent. 119: 587-594

Tallamy, D.W. and Raupp, M.J. 1991. Phytochemical induction by herbivores, John Wiley and Sons, New York.

Tanada, Y. and Kaya, H.K. 1993. Insect Pathology. Academic Press, San Diego, 666 pp.

Thibout, E., Guillot, J.F.and Auger, J. 1993. Microorganisms are involved in the production of volatile kairomones affecting the host seeking behaviour of *Diadromus pulchellus*, a parasitoid of *Acrolepiopsis assectella*. Physiol. Entomol. 18: 176-182.

Thompson, J.N. 1988. Coevolution and alternative hypotheses on insect/plant interactions. Ecology 69: 893-895.

Tietjen, W.J., Ayyagari, L.R. and Uetz, G.W. 1987. Symbiosis between social spider and yeast: the role in prey attraction. Psyche 94: 151-158.

Tieleman, W.C. 1988. Substrate preference of *Leptopilina heterotoma, L. clavipes* and *L. fimbriata*. MSc thesis. Department of Entomology, Wageningen Agricultural University and Department of Population Biology, University of Leiden. (in Dutch).

Turlings, T.C.J., Tumlinson, J.H. and Lewis, W.J. 1990. Exploitation of herbivore-induced plant odors by host-seeking parasitic wasps. Science 250: 1251-1253.

Turlings, T.C.J., Wäckers, F.L., Vet, L.E.M., Lewis, W.J. and Tumlinson, J.H. 1993. Learning of host-finding cues by Hymenopterous parasitoids. In: D.R. Papaj and A.C. Lewis (Eds.) Insect learning pp. 51-78. Chapman and Hall, New York.

Vet, L.E.M. 1985. Olfactory microhabitat location in some eucoilid and alysiine species (Hymenoptera), larval parasitoids of Diptera. Neth. J. Zool. 35: 720-730.

Vet, L.E.M. and Dicke, M. 1992. Ecology of infochemical use by natural enemies in a tritrophic context. Annu. Rev. Entomol. 37: 141-172.

Vinson, S.B. and Iwantsch, G.F. 1980. Host suitability for insect parasitoids. Ann. Rev. Entomol. 25: 397-419.

Vinson, S.B. and Stoltz, D.B. 1986. Cross-protection experiments with two parasitoid (Hymenoptera: Ichneumonidae) viruses. Ann. Entomol. Soc. Am. 79: 216-218.

Visser, J.H. 1986. Host odor perception by phytophagous insects. Annu. Rev. Entomol. 31: 121-144.

Weseloh, R.M. 1981. Host location by parasitoids. In: D.A. Nordlund, R.L. Jones and W.J. Lewis (eds.), Semiochemicals: their role in pest control, p. 79-95. Wiley, New York.

Wit, P.J.G.M. de, 1992. Molecular characterization of gene-for-gene systems in plant-fungus interactions and the application of avirulence genes in control of plant pathogens. Annu. Rev. Phytopathol. 30: 391-418.

Wood, T.G. and Thomas, R.J. 1989. The mutualistic association between Macrotermitinae and *Termitomyces*. In: N. Wilding, N.M. Collings, P.M. Hammond and J.F. Webber (eds.) Insect-Fungus Interactions. pp. 68-92. Acad. Press.

Wood, D.L. 1982. The role of pheromones, kairomones, and allomones in the host selection and colonization behavior of bark beetles. Annu. Rev. Entomol. 27: 411-446.

Zeringue, H.J. and McCormick, S.P. 1989. Relationships between cotton leaf-derived volatiles and growth of *Aspergillus flavus*. J. Am. Oil Chem. Soc. 66: 581-585.

10

SENSITIVITY ANALYSIS IN MICROBIAL COMMUNITIES

H. Nakajima[1] and Z. Kawabata[2]

[1] Department of Physics
Ritsumeikan University
Kusatsu 525, Japan
[2] Department of Environmental Conservation
Ehime University
Matsuyama 790, Japan

A gnotobiotic aquatic microcosm, consisting of three species, including: bacteria, *Escherichia coli*; protozoa, *Tetrahymena thermophila*; and a primary producer, *Euglena gracilis*, was prepared, in which all species co-existed for over 130 days. Interactions among the species were analysed to clarify mechanisms involved in co-existence. The growth of *T. thermophila* depended on the number of *E. coli*, that is, *T. thermophila* grew when more than 10^6 *E. coli* cells per ml were present. No *T. thermophila* survived in a medium without other organisms. It, however, did grow in the presence of *Eu. gracilis*. As *T. thermophila* does not feed on *Eu. gracilis*, it is hypothesized that a metabolite of *Eu. gracilis* had a positive effect on growth of *T. thermophila*. Competition between *E. coli* and *Eu. gracilis* for proteose peptone in the medium was not observed. These interactions appeared to depend on successful co-existence of the three species. These results were confirmed by mathematical analysis, using parameter sensitivity of a dynamic model.

INTRODUCTION

Our primary question was what effects do biological interactions have on the maintenance of species diversity in microbial communities. Several factors sustaining species diversity have been reported: (1) spacial heterogeneity of environments; (2) temporal heterogeneity of environments; and (3) biological interactions among species in the community (Begon et. al., 1986). Many theoretical or mathematical concepts have been proposed for the third factor and tested in natural communities (e.g., Pimm, 1991).

We analyzed relationships between species diversity and biological interactions using an artificial community (a microcosm with neither spacial nor temporal heterogeneities). In defining population interactions in a community, a gnotobiotic microcosm, into which clearly defined species are introduced, is useful because unknown variables are kept to a minimum. Several experiments were done to learn why and/or how successful co-existence

Microbial Diversity in Time and Space, edited by Colwell et al.
Plenum Press, New York, 1996

of the three species occurred. After these experiments, mathematical analysis was used to confirm the reason for the species co-existence obtained in our experiments.

Indirect effects may play an important role in generating the complexity and structural contingency of species interactions in ecological networks (Higashi and Nakajima, 1995). The indirect effects can be estimated by mathematical models, using inflow and parameter sensitivity analysis (Nakajima, 1992; Higashi and Nakajima, 1995; Nakajima and Higashi, 1995). Mathematical analyses also gave us insight into the balance of the strength of biological interactions that is difficult to obtain from experiments alone.

The purpose of this study was: 1) to synthesize a gnotobiotic microcosm in which all species co-exist; 2) to analyze their interactions; and 3) to use mathematical models to learn how co-existence is achieved.

MATERIALS AND METHODS

Organisms: The biotic components of the microcosm were prepared by inoculating the bacteria *Escherichia coli* DH5α, protozoa *Tetrahymena thermophila* B, and algae *Euglena gracilis* Z into a culture medium. The organisms satisfied the following conditions: 1) axenic cultures to avoid introducing unknown organisms; 2) had already been studied, with extensive physiological data available; 3) organisms commonly observed in aquatic systems; 4) except for the bacteria, morphological identification of each organism was possible under the microscope to permit for enumeration; 5) organisms typify trophic levels, such as decomposer, producer, and consumer.

Preparing the microcosms: Ten ml of half strength #36 Taub and Doller solution (1/2 #36 TD) (Taub and Dollar, 1968), without $NaNO_3$, and containing proteose peptone (Difco laboratory, USA) was transferred to test tubes with a screw cap and autoclaved. Into this medium, organisms were inoculated aseptically. The concentrations of proteose peptone employed were 0.0, 10, 50, 100, 500 and 1000 mg per l of 1/2 #36 TD. Test tubes containing culture media and organisms were placed in an incubator, fitted with fluorescent lamps, under conditions of 2500 lux and a 12-12 LD light regime, at 25°C.

Enumeration: Colonies on a broth-agar medium incubated at 25°C for 5 days in a petridish were counted for culturable *E. coli*. The colonies were counted in triplicate platings and the mean value of *E. coli* per ml was calculated. One ml of sample was transferred to a counting chamber 1 mm deep, 50 x 20 mm^2 and meshed in 1 mm^2 sector. All the cells of *T. thermophila* in ten 1 mm^2 areas, randomly selected, were counted microscopically. The counts, which were repeated 5 times using the same sample, yielded a 2.5% coefficient deviation of s/<x> x 100 (s= standard deviation of the sample and <x>= mean value of sample). *Eu. gracilis* was counted in the same way as *T. thermophila*.

RESULTS

Population changes of each organism in mixed culture of the three species in a culture medium containing the various concentrations of proteose peptone are shown in Fig. 1. Co-existence of the three species was attained for more than 130 days in the medium containing more than 50 mg of proteose peptone per l l of 1/2 #36 TD.

Several cultures of an individual or a pair of species in 0.05% proteose peptone culture medium were studied to analyze interactions among the species. The population changes for the individual species, *E. coli* and *T. thermophila* and for a mixed culture of those species showed that *T. thermophila* grew by predation on *E. coli* (Fig. 2).

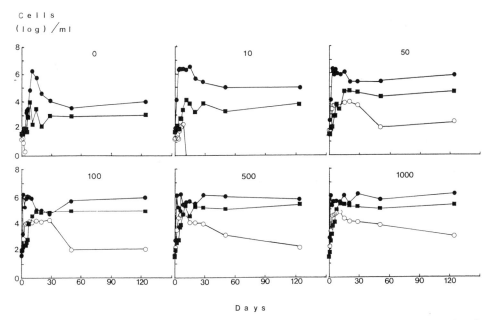

Figure 1. Population changes of each organism (●: *E. coli*, ○: *T. thermophila*, ■: *Eu. gracilis*). Numbers in the graph indicate concentrations of proteose peptone (mg l-1).

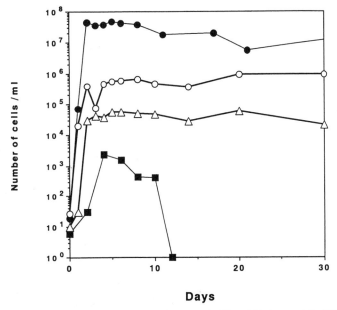

Figure 2. Population changes for an individual species, *E. coli* (●) and *T. thermophila* (■) and for a mixed culture of those species, *E. coli* with *T. thermophila* (○) and *T. thermophila* with *E. coli* (△).

However, predation depended on the density of prey, as shown in Figs. 3 and 4. No predation was observed when the density of *E. coli* was less than 10^6 cells ml^{-1}. Population changes for the individual species, *Eu. gracilis* and *T. thermophila*, and for a mixed culture of those species, showed that *T. thermophila* grew with increased numbers of *Eu. gracilis*, although *T. thermophila* died in pure culture alone (Fig. 5). It appears that a metabolite of *Eu. gracilis* had a positive effect on growth of *T. thermophila*. Population changes for an individual species alone, *E. coli* or *Eu. gracilis*, and for a mixed culture of those species showed that both grew similarly in both single and mixed cultures (Fig. 6). There was no competition observed between *E. coli* and *Eu. gracilis* for organic nutrient, that could be discerned by these experiments.

It is possible that successful co-existence of these three species was a function, not only of the conditions of the culture medium, but also of the biological interactions.

MATHEMATICAL ANALYSIS

Initially, mathematical analysis, using a dynamic model of the three species system, without competition between the bacteria and algae was employed:

$$dE/dt = [\varepsilon_1 - E - aE^2T/(h + E^20)]E$$
$$dT/dt = [-\varepsilon_2 + cE^3/(h + E^2) + rU/(k + T)]T$$
$$dU/dt = (\varepsilon_3 - U)U \qquad (1)$$

where E, T and U are concentrations of *E. coli*, *T. thermophila* and *Eu. gracilis*, respectively.

In this model, a switching of predation is considered, which is represented by $aE^2/(h+E^2)$ in the predation coefficient. When $h = 0$, there is no switching, so the parameter h means the strength of the predation switching. A parameter r represents the efficiency of metabolite utility by *T. thermophila*. When $r = 0$, *T. thermophila* does not use the *Eu. gracilis* metabolite.

In the case where there is neither switching of predation nor metabolite utilization by *T. thermophila*, the conditions of the three species co-existence is represented by inequality,

$$\varepsilon_1 > \varepsilon_2/c . \qquad (2)$$

When inequality (2) is not satisfied, *T. thermophila* becomes extinct. The intrinsic growth rate of *E. coli* must be greater than a certain level for *T. thermophila* to survive.

We calculated the changes in the steady state level of the three species caused by parameter changes, i.e., dE/dp, dT/dp and dU/dp, where p is a parameter and E, T and U are the steady state levels of the species. We obtained parameter sensitivity, with respect to efficiency of metabolite utilization, as follows,

$$dE/dr = -aE^2U/[\Gamma(h + E^2)(k + T)] < 0,$$
$$dT/dr = [1 + 2ahET/(h + E^2)^2]U/\Gamma(k + T)] > 0,$$
$$dU/dr = 0. \qquad (3)$$

From the second equation of eq. (3), we found that, when the efficiency of metabolite utilization increases, the steady state level of *T. thermophila* also increases. Thus, a given level of metabolite utilization may rescue *T. thermophila* from extinction, even when inequality (2) does not hold, i.e., *T. thermophila* cannot survive without metabolite utiliza-

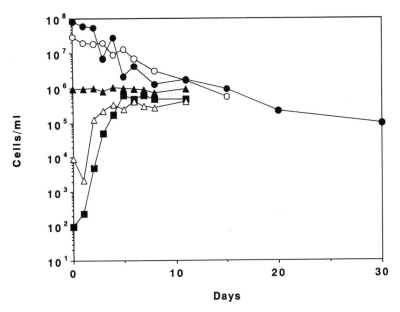

Figure 3. Population changes in *E. coli* with different initial concentrations and with predation by *T. thermo-phila*.

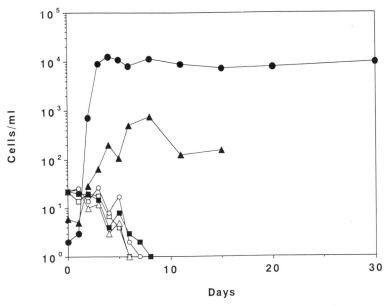

Figure 4. Growth of *T. thermophila* with various concentrations of *E. coli* ($\square$: without *E. coli*, $\triangle$: 10^2 cells ml^{-1}, $\bigcirc$: 10^4 cells ml^{-1}, $\blacksquare$: 10^6 cells ml^{-1}, $\blacktriangle$: 10^7 cells ml^{-1}, $\bullet$:10^8 cells ml^{-1}).

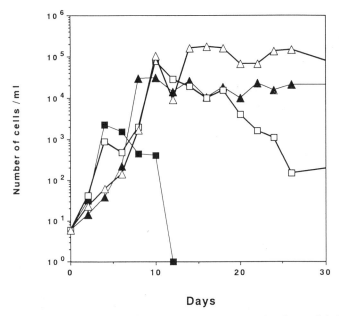

Figure 5. Population changes for an individual species, *Eu. gracilis* (▲) and *T. thermophila* (■) and for mixed culture of those species, *Eu. gracilis* with *T. thermophila* (△) and *T. thermophila* with *Eu. gracilis* (□).

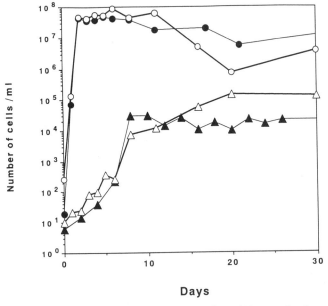

Figure 6. Population changes for the species, *E. coli* alone (●) and *Eu. gracilis* alone (▲) and for a mixed culture of those species, *E. coli* with *Eu. gracilis* (○) and *Eu. gracilis* with *E. coli* (△).

goes extinct when r increases. Therefore, we obtained the following result: if there is a switching predation (concentration dependent predation), then increased metabolite utilization relaxes the restriction of the co-existence of the three species.

From the parameter sensitivity, with respect to the intensity of predation switching which can be obtained as in the case of change in metabolite utilization, we obtained the following results: (a) If a system consisting of *T. thermophila* and *Eu. gracilis* survives without *E. coli*, then a system with the three species can also survive at arbitrary switching levels of predation; (b) if *T. thermophila* cannot survive without *E. coli*, then *T. thermophila* becomes extinct when the predation switching level is high.

When there is competition between *E. coli* and *Eu. gracilis*, the model (1) is modified as follows,

$$dE/dt = [\varepsilon_1 - E - aE^2T/(h + E^2) - pU]E$$
$$dT/dt = [-\varepsilon_2 + cE^3/(h + E^2) + rU/(k + T)]T$$
$$dU/dt + (\varepsilon_3 - U - qE)U, \tag{4}$$

where p and q are the competition coefficients.

From the mathematical analysis of eq. (4), using parameter sensitivity with respect to the competition coefficients, we obtained the result that, since competition between bacteria and algae reduced the steady state level of the protozoa, the competition reduced the parameter region in which the three species could co-exist.

ACKNOWLEDGMENT

This work was supported, in part, by a Japan Ministry of Education, Science and Culture Grant-in-Aid for Scientific Research on Priority Areas (#319), Project "Symbiotic Biosphere: An Ecological Interaction Network Promoting the Coexistence of Many Species".

REFERENCES

Begon, M., Harper, J. L., and Townsend, C. R. 1986. Ecology: Individuals, Population and Communities, Blackwell Scientific Publication.

Higashi, M. and Nakajima, H. 1995. Indirect Effects in Ecological Interaction Networks I: The Chain Rule Approach. Math. Biosciences, (in press).

Nakajima, H. 1991. Sensitivity and Stability of Flow Networks. Ecol. Modelling, 62:123-133.

Nakajima, H. and Higashi, M. 1995. Indirect Effects in Ecological Interaction Networks II: The Conjugate Variable Approach. Math. Biosciences, (in press).

Pimm, S. L. 1991. Balance of Nature, The University of Chicago Press, Chicago IL.

Taub, F. B. and Dollar, A. M. 1968. The Nutritional Inadequancy of Chlorella and Chlamydomonas as Food Daphnia pulex. Limnol. Oceanogr, 13: 607-617.

11

INTRACELLULAR SYMBIOSIS IN INSECTS

Hajime Ishikawa

Department of Biological Sciences
Graduate School of Science
University of Tokyo
Hongo, Bunkyo-ku, Tokyo 113, Japan

INTRODUCTION

Many eukaryotic cells constitute the sole habitat for a vast and varied array of prokaryotic lineages (Buchner, 1965). These intracellular associations have evolved repeatedly and have had major consequences for the diversification of both bacteria and host. The magnitude of these consequences is immediately evident if one considers the examples of mitochondria and chloroplasts, now widely acknowledged to be descended from prokaryotes that invaded intracellular habitats (Margulis, 1970; Gray and Doolittle, 1982).

A considerable number of insect species harbor intracellular symbionts that are vertically transmitted through host generations. These symbionts are classified into two different types. One is the so-called mycetocyte symbionts, which are harbored by the host mycetocyte, or bacteriocyte, a huge cell differentiated for this purpose (Ishikawa, 1989). In general, the host and its mycetocyte symbionts are closely mutualistic and indispensable to each other for their growth and reproduction. Host provides symbionts with metabolic substrates, as well as a safe and stable niche, while symbionts benefit the host by producing organic compounds, such as several amino acids that are not readily available to the host (Douglas, 1989; Sasaki et al., 1990). Mycetocyte symbionts, when housed in the host cell, selectively synthesize symbionin, a stress protein homologous to *E. coli* GroEL (Hara et al., 1990). It has been suggested that symbionin not only functions as molecular chaperone (Kakeda and Ishikawa, 1991; Ohtaka et al., 1992) but also is a histidine protein kinase, serving as a 'sensor' molecule of the two-component pathway of signal transduction (Stock et al., 1989; Morioka et al., 1993; 1994).

The other type of intracellular symbiont is called "guest microbe", whose habitat is not restricted to a particular cell type, such as the mycetocyte, but is present in almost all the cell types of the host insect. Guest microbes are not apparently mutualistic with their host, but exploit the host by interfering with its sexuality and reproduction (Hoffman et al., 1986; Breeuwer and Werren, 1990). In this article, I survey two types of symbiosis, from the evolutionary point of view.

Microbial Diversity in Time and Space, edited by Colwell et al.
Plenum Press, New York, 1996

MYCETOCYTE SYMBIOSIS

In view of the location of the mycetocyte symbiont, it is generally believed that these symbionts originated from gut microbes which are descendants of free-living bacteria taken in with diets by the insects in the evolutionary past (Buchner, 1965). Actually, recently we demonstrated that an aphid mycetocyte symbiont is more closely related to a gut microbe of the same aphid, in terms of the nucleotide sequences of 16S rRNA gene (16S rDNA) and other genes (Harada and Ishikawa, 1993). While this gut microbe was a close relative of *E. coli*, the mycetocyte symbiont was closer to this microbe than to *E. coli*. In addition, the optimal temperature for the growth of this microbe was significantly lower than that of *E. coli*, suggesting that the microbe adapted to the ambient temperature of the host insect and shares a very close ancestor with the mycetocyte symbiont. Interestingly, an aseptic host, deprived of the gut microbe, showed better performance, in terms of both growth and fecundity. This suggests that once the host establishes an intracellular symbiosis with a microbe, its descendants in the gut are nothing but a nuisance to the host (Harada, unpublished data).

Members of at least six insect orders harbor mycetocyte symbionts, which are, in many cases, bacteria, but sometimes yeast-like eukaryotes may be present (Buchner, 1965). These symbionts benefit their hosts inasmuch as hosts treated with antibiotics cease ordinary growth patterns and usually die without producing progeny. That the symbionts also benefit from the association is usually assumed because their habitat is restricted to the insect cells (Ishikawa, 1989; Douglas, 1989). The mycetocyte bacteria of many insect taxa are not well studied, and because the associations seem diverse in form and function, it is difficult to generalize about them. Nevertheless, much has been learned from aphids and other homopterous insects that feed on plant sap.

To summarize:

i) insect offspring do not acquire mycetocyte endosymbionts from the environment or from food; they usually inherit symbionts maternally via transovarial transmission;

ii) apparently the main benefit conferred upon the host by symbionts is nutritional augmentation by mechanisms such as recycling uric acid N (in case of cockroaches; Cochran, 1985) or glutamine N (in case of aphids; Sasaki and Ishikawa, 1993; Sasaki et al., 1993; Sasaki and Ishikawa, 1995) in order to synthesize host-essential amino acids;

iii) these mycetocyte bacteria have resisted cultivation, despite many attempts, although they can be mechanically isolated, maintained, and studied for a short period in a metabolically active state (Ishikawa, 1984; Whitehead and Douglas, 1993).

Morphological differences between prokaryotic symbionts in mycetocytes from different insect taxa, or sometimes within a host, and differences in location of mycetocytes (e.g., in a fat body, free in the haemocoel or associated with the digestive tract; Buchner, 1965) suggest that the prokaryotic mycetocyte symbionts are probably polyphyletic. Comparative sequence analysis of 16S rDNA revealed that mycetocyte bacteria appear to be specialized, with respect to their hosts (Baumann et al., 1993). For example, among sternorrhynchous homopterans, aphids, whiteflies and mealybugs, each harbor their own lineage of mycetocyte bacteria. Those from aphids and whiteflies are members of the gamma-subdivision of the Proteobacteria, while those from mealybugs are affiliated with the beta-subdivision, and each lineage is distinguished from other lineages within the Proteobacteria (Munson et al., 1992; Clark et al., 1992).

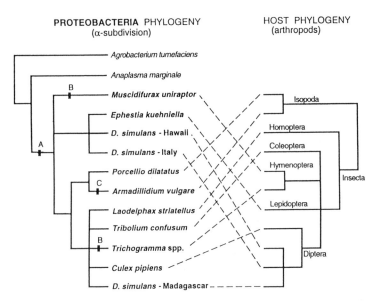

PROTEOBACTERIA PHYLOGENY
(α-subdivision)

HOST PHYLOGENY
(arthropods)

Figure 1. Phylogeny of the mycetocyte symbionts of aphids and phylogeny of the corresponding aphid hosts. The bacterial phylogeny is based on 16S rDNA sequences; the aphid phylogeny is based on morphology. Dashed lines indicate associations. Taxa within mycetocyte symbionts are represented by names of their aphid hosts. Dates are estimated from aphid fossils and/or biogeography; they apply to both aphid and bacterial ancestors (Moran and Baumann, 1994, with permission).

The mycetocyte symbioses of aphids are the best studied, both in terms of the nature of the interactions (Ishikawa et al., 1992) and their evolutionary histories, as revealed by molecular phylogenetics (Moran et al., 1993). Currently, the 16S rDNA sequences of mycetocyte symbionts of 12 aphid species have been obtained and analyzed phylogenetically together with sequences of other representative prokaryotes. Results indicate that all aphid primary symbionts, the inhabitants of the large mycetocytes, belong to a single, well-supported clade within the gamma-3 subgroup and have *E. coli* and related organisms as their closest relatives (Moran et al., 1993). The formal designation, *Buchnera aphidicola*, applies to this symbiont clade (Munson et al., 1991). It turned out that the sequence-based phylogeny of *Buchnera* is completely concordant with the morphology-based phylogeny of the corresponding aphid hosts. Since the probability of such concordance occurring by chance is diminishingly small, these results imply a single original infection in a common ancestor of the included aphid, followed by cospeciation of aphids and *Buchnera*. The distribution of modern symbionts among their host aphids thus reflects parallel cladogenesis through consistent, long-term vertical transmission from mother to daughter. Horizontal transfer of these symbionts appears to be rare or absent (Moran et al., 1993; Moran and Baumann, 1994). The aphid fossil record implies that the symbiotic association is ancient. Several of the descendant lineages are known from an 80 million year old amber deposit, implying a minimum age for the most recent common ancestor of this set of aphids as well as for the date of the original infection (Heie, 1987). In fact, comparative sequence analyses of 16S rDNA suggested that the common ancestor is considerably older, its age having been estimated at about 200-250 million years. Also, *Buchnera* has been shown to be distant from any previously sequenced bacteria, an observation consistent with an ancient origin of the association followed by divergence of the symbiotic lineage from other bacteria (Moran and Baumann, 1994).

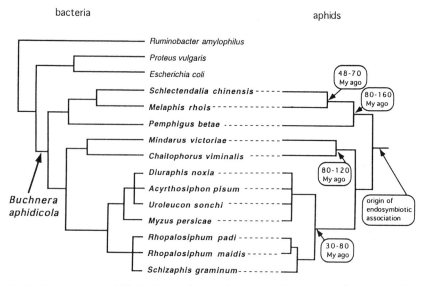

Figure 2. Phylogeny of selected Wolbachia species and phylogeny of the corresponding arthropod hosts. The bacterial phylogeny is compiled from anlyses based on 16S rDNA sequences. Effects of Wolbachia infection: A, cytoplasmic incompatibility; B, parthenogenesis; C, ferminization (Moran and Baumann, 1994, with permission

The radiation of the Homoptera into the sap-feeding niches provided by vascular plants was presumably dependent on the early acquisition of bacterial mutualists. This possibility is supported by the fact that most such diets lack nutrients essential for insects and that almost all modern Homoptera contain symbionts (Buchner, 1965; Douglas, 1989). These considerations suggest that the infection that led to modern *Buchnera* might have occurred in an ancestor common to aphids and related insects. However, sequence analysis has indicated that symbionts have been acquired more than once within the sternorrhynchous Homopterae.

It has been known that many insects, particularly many homopterans, contain more than one type of symbiont, confined to separate host cells (Buchner, 1965; Hinde, 1971). Sequence analysis indicated that secondary symbionts in certain aphids are distinct from *Buchnera* and arose from an independent infection by another member of the gamma-3 subgroup (Unterman et al., 1989; Munson et al., 1991). It suggested also that aphid secondary symbionts comprize a diverse assemblage of bacterial types, in contrast to the uniformity of the primary symbionts mentioned above (Moran and Baumann, 1994). Immunohistochemistry and immunoblot analysis using antiserum against a symbiont-specific stress protein indicated that acquisition of secondary symbionts by the host is much more recent than that of *Buchnera* (Fukatsu & Ishikawa, 1993). The location and presence of secondary symbionts within an aphid species appear to vary much more than for *Buchnera*, adding to evidence that secondary association in aphids are relatively labile. Analysis of 16S rDNA sequence for the secondary symbiont of a whitefly also indicates an additional ancestral infection.

While mycetocyte-inhabiting prokaryotic symbionts are conserved amongst aphids, evidence suggests replacement of a symbiont by another microbe in the case of certain members of the aphid tribe Cerataphidini (Fukatsu and Ishikawa, 1992). These species lack *Buchnera* but contain yeast-like organisms that live free in the haemocoel. Since these aphid species are related and are nested within the clade of aphids containing *Buchnera*, they appear to be a clade descended from a common ancestor in which *Buchnera* was replaced with an extracellular eukaryotic associate (Fukatsu et al., 1994). The 18S rDNA sequence suggested that the extracellular symbiont is a phylogenetical relative of *Neurospora* that belongs to Pyrenomycetidae (Fukatsu, unpublished data). It is an interesting possibility that the symbiont stems from an entomogenous fungus. Circumstantial evidence suggests that the replacement of the symbiont enabled the group of aphids to feed on the diets that they could otherwise not utilize.

GUEST MICROBES

In contrast to mycetocyte-inhabiting symbionts so far described, some microorganisms, such as *Wolbachia*, have effects that enhance their own spread but apparently have zero or negative effect on host fitness. They are effectively parasitic, although the severity of their effects varies. In these instances, in which effects of an association are not known, bacterial associates are referred to simply as maternally or cytoplasmically inherited microorganisms, which are so-called guest microbes (Ebbert, 1993).

One type of guest microbe was detected in conjunction with two phenomena related to insect reproduction and sexuality. The first of these, cytoplasmic incompatibility (Hoffman et al., 1986; Breeuwer and Werren, 1990), was originally thought to be a relatively unusual phenomenon found in a few insect species. This incompatibility results from bacteria that interfere with sperm-derived chromosomes in fertilized eggs by preventing proper condensation (O'Neill and Karr, 1990). A typical consequence is that symbiont-containing males crossed with symbiont-free females produce no viable progeny. However, the reciprocal cross, i.e., cross between symbiont-free males and symbiont-containing females, produces normal numbers of progeny, as do crosses in which both parents contain symbionts. The second case is when guest microbes cause parthenogenesis by preventing segregation of chromosomes in unfertilized eggs (Stouthamer et al., 1993). In the parasitoid wasps, *Trichogramma* and *Muscidifurax*, for example, this results in thelytoky, or the production of only female progeny (Stouthamer et al., 1990; Stouthamer and Luck, 1993). It is beyond question that symbiont-mediated effects, such as cytoplasmic incompatibility and parthenogenesis, can have a strong impact on the population dynamics and evolution of insects (Ebbert, 1993).

Recently, 16S rDNAs of bacteria responsible for cytoplasmic incompatibility in mosquitos, beetles, wasps, a moth and a fruit fly were sequenced (O'Neill et al., 1992; Rousset et al., 1992; Breeuwer et al., 1992; Stouthamer et al., 1993). As a result, it turned out that cytoplasmic incompatibility-causing bacteria surveyed to date are closely related to one another, belonging to the alpha-subgroup of the Proteobacteria; they are all different species or strains of the genus *Wolbachia* (Hertig, 1936). As in mycetocyte symbionts, *Wolbachia* infections are maternally inherited through transmission from the mother to her eggs before oviposition. Regular maternal inheritance would seem to create the potential for cospeciation of *Wolbachia* and its hosts. However, the phylogenetic findings based on *Wolbachia* sequences (O'Neill et al., 1992; Rousset et al., 1992) provide an interesting contrast to the findings for *Buchnera*. Results indicate that *Wolbachia* lineages have sometimes been transmitted among members of different insect orders (Moran and Baumann, 1994). The amount of divergence among the most distant of the examined *Wolbachia* 16S sequences is small (<3%) (O'Neill et al., 1992) relative to that among distantly related *Buchnera* (>8%) (Moran et al., 1993). Provided that substitution rates in *Wolbachia* are not much different from those estimated for *Buchnera* and other prokaryotes, the common ancestor of examined *Wolbachia* lived about 25-100 million years ago. This is much younger than any common ancestor of the hosts (>300 million years), adding to evidence that the distribution of *Wolbachia* is the result, in part, of occasional horizontal transfer of bacterial lineages among lineages of arthropods (Moran and Baumann, 1994). Evidence for more than one *Wolbachia* strain within individual insects (Rousset et al., 1992) is consistent with the view that horizontal transfer has been an important determinant of the modern distribution of *Wolbachia* among hosts.

Sterility in crosses between *Wolbachia*-infected males and uninfected females suggests the infection interferes with the function of sperm or the male chromosomes and that an infected oocyte is necessary to rescue proper function of the male gamete (Yen and Barr, 1971). In some insect species, chromosomes from the sperm of infected males fail to condense in the cytoplasm of an uninfected egg, but do condense in that of an infected egg (Breeuwer and Werren, 1990; O'Neill and Karr, 1990; Montchamp-Moreau, 1991). In this context, incompatibility infections in *Culex* mosquitoes are intriguing, in terms of speciation of this insect genus. In *Culex* mosquitoes, no naturally occurring uninfected individuals have been reported. Matings between infected individuals typically succeed if both individuals derive from the same population, yet can fail in crosses between infected individuals from distinct populations. These observations can be explained by assuming that there are variants of *Wolbachia* due to host population and that only the cytoplasm of the egg infected with the same variant is able to rescue the chromosomes from infected sperm (Ebbert, 1993).

Wolbachia are expected to spread rapidly in an infected population and presumably, between populations that experience significant gene flow. *Wolbachia* will sabotage the matings of uninfected females, promoting the fitness of their own hosts and the spread of the infection. Once the infection becomes fixed in a population, variation in *Wolbachia* populations isolated in maternal lineages is expected and can generate further incompatibilities between infected individuals. Assuming these variants spread, interpopulation incompatibility may occur; this is the basis for the suggestion that these infections can drive speciation (Ebbert, 1993).

Although guest microbes, exemplified by *Wolbachia*, seem to be typically selfish in the present day, there is some possibility that they will evolutionarily change themselves to mutualists like *Buchnera*. Since the propagation of guests is dependent, at least in part, on host fecundity, selection on both hosts and bacteria will favor any traits that enhance the contribution of the bacteria to host fitness, potentially resulting in a mutually beneficial interactions. Such adaptive evolution could underlie the eventual sequestration of the bacteria within certain specialized cells as well as special mechanisms for their transfer from mother to offspring (Thompson, 1982).

CONCLUSIONS

The effects of cytoplasmically inherited microbes range from those of obligate mutualists required for host survival to those of selfish parasites that may cause all of a host's progeny to die prematurely. The great diversity of insects is partly due to their frequent associations with mutualistic endosymbionts, such as *Buchnera*, which allow hosts to exploit niches that would otherwise be nutritionally unsuitable. In other instances, bacterial associates may promote speciation through direct effects on reproductive systems. For example, induction of incompatibility or parthenogenesis by *Wolbachia* could effect reproductive isolation of host populations (Stouthamer et al., 1990). The extent of mutualism versus antagonism underlying a particular association will determine the degree of conflict between selection pressures on host and prokaryote, with implications for the evolution of both participating lineages (Moran and Baumann, 1994).

REFERENCES

Baumann, P., Munson, M. A., Lai, C. -Y., Clark, M. A., Baumann, L., Moran, N. A., and Campbell, B. C., 1993, Origin and properties of bacterial endosymbionts of aphids, whiteflies, and mealybugs, *ASM News* *59*: 21-24.

Breeuwer, J. A. J., and Werren, J. H., 1990, Microorganisms associated with chromosome destruction and reproductive isolation between two insect species, *Nature 346*: 558-560.

Breeuwer, J. A. J., Stouthamer, R., Barns, S. M., Pelletier, D. A., Weisburg, W. G., and Werren, J. H., 1992, Phylogeny of cytoplasmic incompatibility microorganisms in the parasitoid wasp genus *Nasonia* (Hymenoptera: pteromalidae) based on 16S ribosomal DNA sequences, *Insect Mol. Biol. 1*: 25-36.

Buchner, P., 1965, *Endosymbiosis of Animals with Plant Microorganisms,* Interscience Publishers, New York.

Clark, M. A., Baumann, L., Munson, M. A., Baumann, P., Campbell, B. C., Duffus, J. E., Osborne, L. S., and Moran, N. A., 1992, The eubacterial endosymbionts of whiteflies (Homoptera: Aleyrodoidea) constitute a lineage distinct from the endosymbionts of aphids and mealybugs, *Curr. Microbiol. 25*: 119-123.

Cochran, D. G., 1985, Nitrogen excretion in cockroaches, *Ann. Rev. Entomol. 30*: 29-49.

Douglas, A. E., 1989, Mycetocyte symbiosis in insects, *Biol. Rev. 64*: 409-434.

Ebbert, M. A., 1993, Endosymbiotic sex ratio distorters in insects and mites, in *Evolution and Diversity of Sex Ratio in Insects and Mites* (ed. Wrensch, D. L. and Ebbert, M. A.), Chapman & Hall, New York.

Fukatsu, T., and Ishikawa, H., 1992, A novel eukaryotic extracellular symbiont in an aphid, *Astegopteryx styraci* (Homoptera, Aphididae, Hormaphidinae), *J. Insect Physiol. 38*: 765-773.

Fukatsu, T., and Ishikawa, H., 1993, Occurrence of chaperonin 60 and chaperonin 10 in primary and secondary symbiont of aphids: Implications for evolution of endosymbiotic system in aphids, *J. Mol. Evol. 36*: 568-577.

Fukatsu, T., Aoki, S., Kurosu, U., and Ishikawa, H., 1994, Phylogeny of Cerataphidini aphids revealed by their symbiotic microorganisms and basic structure of their galls: Implications for host-symbiont coevolution and evolution of sterile soldier castes, *Zoological Sci. 11*: 613-623.

Gray, M. W., and Doolittle, W. F., 1982, Has the endosymbiont hypothesis been proven ? *Microbiol. Rev. 46*: 1-42.

Hara, E., Fukatsu, T., Kakeda, K., Kengaku, M., Ohtaka, C., and Ishikawa, H., 1990, The predominant protein in an aphid endosymbiont is homologous to an *E. coli* heat shock protein, *Symbiosis 8*: 271-283.

Harada, H., and Ishikawa, H., 1993, Gut microbe of aphid closely related to its intracellular symbiont, *BioSystems 31*: 185-191.

Heie, O. E., 1987, Paleontology and phylogeny, in *Aphids: Their Biology, Natural Enemies and Control* Vol. 2A (ed. Minks, A. K. and Harrewijn, P.), Elsevier, Amsterdam, pp. 367-391.

Hertig, M., 1936, The rickettsia *Wolbachia pipiens* (gen. et sp. n.) and associated inclusions of the mosquito, *Culex pipiens, Parasitol. 28*: 453-486.

Hinde, R., 1971, The fine structure of the mycetocyte symbiotes of the aphids *Brevicoryne brassicae, Myzus persicae* and *Macrosiphum rosae, J. Insect Physiol. 17*: 2035-2050.

Hoffmann, A. A., Turelli, M., and Simmons, G. M., 1986, Unidirectional incompatibility between population of *Drosophila simulans, Evolution 40*: 692-701.

Ishikawa, H., 1984, Characterization of the protein species synthesized *in vivo* and *in vitro* by an aphid endosymbiont, *Insect Biochem. 14*: 417-425.

Ishikawa, H., 1989, Biochemical and molecular aspects of endosymbiosis in insects, *Intern. Rev. Cytol. 116*: 1-45.

Ishikawa, H., Fukatsu, T., and Ohtaka-Maruyama, C., 1992, Cellular and molecular evolution of intracellular symbiont, in *The Origin and Evolution of the Cell* (ed. Hartman, H. and Matsuno, K.), World Scientific, Singapore, pp. 205-229.

Kakeda, K., and Ishikawa, H., 1991, Molecular chaperone produced by an intracellular symbiont, *J. Biochem. 110*: 583-587.

Margulis, L., 1970, *Origin of Eukaryotic Cells*, Yale University Press, New Haven.

Montchamp-Moreau, C., Ferveur, J. F., and Jacques, M., 1991, Geographic distribution and inheritance of three cytoplasmic incompatibility types in *Drosophila simulans, Genetics 129*: 399-407.

Moran, N. A., Munson, M. A., Baumann, P., and Ishikawa, H., 1993, A molecular clock in endosymbiotic bacteria is calibrated using insect host, *Proc. R. Soc. Lond. B 253*: 167-171.

Moran, N. A., and Baumann, P., 1994, Phylogenetics of cytoplasmically inherited microorganisms of arthropods, *Trends Ecol. Evol. 9*: 15-20.

Morioka, M., Muraoka, H., and Ishikawa, H., 1993, Chaperonin produced by an intracellular symbiont is an energy-coupling protein with phosphotransferase activity, *J. Biochem. 114*: 246-250.

Morioka, M., Muraoka, H., Yamamoto, K., and Ishikawa, H., 1994, An endosymbiont chaperonin is a novel type of histidine protein kinase, *J. Biochem. 116*: 1075-1081.

Munson, M. A., Baumann, P., and Kinsey, M. G., 1991, *Buchnera* gen. nov. and *Buchnera aphidicola* sp. nov., a taxon consisting of the mycetocyte associated primary endosymbionts of aphids, *Int. J. Syst. Bact. 41*: 566-568.

Munson, M. A., Baumann, P., and Moran, N. A., 1992, Phylogenetic relationships of the endosymbionts of mealybugs (Homoptera: Pseudococcidae) based on 16S rDNA sequences, *Mol. Phylogen. Evol. 1*: 26-30.

Ohtaka, C., Nakamura, H., and Ishikawa, H., 1992, Structure of chaperonins from an intracellular symbiont and their functional expression in *E. coli groE* mutants, *J. Bacteriol. 174*: 1869-1874.

O'Neill, S. L., and Karr, T. L., 1990, Bidirectional incompatibility between conspecific populations of *Drosophila simulans, Nature 348*: 178-180.

O'Neill, S. L., Giordano, R., Colbert, A. M. E., Karr, T. L., and Robertson, H. M., 1992, 16S rRNA phylogenetic analysis of the bacterial endosymbionts associated with cytoplasmic incompatibility in insects, *Proc. Natl. Acad. Sci. USA 89*: 2699-2702.

Rousset, F., Bouchon, D., Pintureau, B., Juchault, J., and Solignac, M., 1992, *Wolbachia* endosymbionts responsible for various alterations of sexuality in arthropods, *Proc. R. Soc. Lond. B 250*: 91-98.

Sasaki, T., Aoki, T., Hayashi, H., and Ishikawa, H., 1990, Amino acid composition of the honeydew of symbiotic and aposymbiotic pea aphids *Acyrthosiphon pisum, J. Insect Physiol. 36*: 35-40.

Sasaki, T., and Ishikawa, H., 1993, Nitrogen recycling in the endosymbiotic system of the pea aphid, *Acyrthosiphon pisum, Zoological Sci. 10*: 779-785.

Sasaki, T., Fukuchi, N., and Ishikawa, H., 1993, Amino acid flow through aphid and its symbiont: Studies with ^{15}N-labeled glutamine, *Zoological Sci. 10*: 787-791.

Sasaki, T., and Ishikawa, H., 1995, Production of essential amino acids from glutamate by mycetocyte symbionts of the pea aphid, *Acyrthosiphon pisum, J. Insect Physiol.*, 41: 41-46.

Stock, J. B., Ninfa, A. J., and Stock, A. M., 1989, Protein phosphorylation and regulation of adaptive responses in bacteria, *Micribiol. Rev. 53*: 450-490.

Stouthamer, R., Luck, R. F., and Hamilton, W. D., 1990, Antibiotics cause parthenogenetic *Trichogramma* (Hymenoptera /Trichogrammatidae) to revert to sex, *Proc. Natl. Acad. Sci. USA 87*: 2424-2427.

Stouthamer, R., and Luck, R. F., 1993, Influence of microbe-associated parthenogenesis on the fecundity of *Trichogramma deion* and *T. pretiosum, Entomol. Exp. Appl. 67*: 183-192.

Stouthamer, R., Breeuwer, J. A. J., Luck, R. F., and Werren, J. H., 1993, Molecular identification of microorganisms associated with parthenogenesis, *Nature 361*: 66-68.

Thompson, J. N., 1982, *Interaction and Coevolution*, John Wiley and Sons, New York.

Unterman, B. M., Baumann, P., and McLean, D. L., 1989, Pea aphid symbiont relationships established by analysis of 16S rRNAs, *J. Bacteriol. 171*: 2970-2974.

Whitehead, L. F., and Douglas, A. E., 1993, A metabolic study of *Buchnera*, the intracellular bacterial symbionts of the pea aphid *Acyrthosiphon pisum, J. Gen. Microbiol. 139*: 821-826.

Yen, J. H., and Barr, A. R., 1971, New hypothesis of the cause of cytoplasmic incompatibility in *Culex pipiens, Nature 232:* 657-658.

12

EVOLUTIONARY RELATIONSHIPS OF CATABOLIC FUNCTIONS IN SOIL BACTERIA

Kensuke Furukawa, Nobutada Kimura, and Jun Hirose

Department of Agricultural Chemistry
Kyushu University
Hakozaki, Fukuoka 812, Japan

INTRODUCTION

Within the microbial world, soil bacteria demonstrate a unique metabolic versatility for degradation of a variety of aromatic compounds. The major aromatic pathways discovered in bacteria revealed that essentially all compounds are degraded through a variety of enzymatic steps to limited numbers of common intermediates, such as catechols, which are key compounds for further metabolism. The relationships among the different aromatic pathways and gene clusters often reveal evolutionary changes involved in the development of metabolic routes (van der Meer *et al.*, 1992). Such evolution derived from various genetic events. Biphenyl-utilizing bacteria, widely distributed in the natural environment, include both Gram negative and Gram positive strains (Furukawa, 1982). The substrate specificities of biphenyl catabolic enzymes are usually very relaxed. Some *bph* genes (coding for biphenyl catabolism) are very similar in different strains, suggesting that certain *bph* genes may transfer among soil bacteria (Furukawa *et al.*, 1989). On the other hand, other *bph* genes are highly diversified and greatly rearranged. Toluene-utilizing bacteria are also widely distributed. Toluene can be metabolized by bacteria by different mechanisms, where substituted groups are modified before or after ring-cleavage, depending on the microorganism. In this communication, evolutionary relationships of catabolic function is discussed, focusing on biphenyl-utilizing and toluene-utilizing bacteria.

BIOCHEMISTRY AND GENETICS OF BIPHENYL METABOLISM IN SOIL BACTERIA

Biphenyl-utilizing bacteria isolated to date are mostly aerobic Gram negative bacteria, which include species of *Pseudomonas, Acinetobacter, Achromobaccer, Alcaligenes,* and *Moraxella*. Gram-positive strains, such as *Arthrobacter* and *Rhodococcus* have also been isolated. These organisms usually metabolize biphenyl via the oxidative route as shown in

Microbial Diversity in Time and Space, edited by Colwell et al.
Plenum Press, New York, 1996

101

Fig. 1. Biphenyl dioxygenase catalyzes the initial oxidation of the biphenyl molecule to the dihydrodiol by introducing two atoms of oxygen, and is composed of terminal dioxygenase, ferredoxin and ferredoxin reductase. The terminal dioxygenase consists of a large subunit (an iron-sulfur protein, encoded by *bphA1* gene) and a small subunit (*bphA2*). The ferredoxin (*bphA3*) and the ferredoxin reductase (*bphA4*) mediate electron transport from NADH to the terminal dioxygenase. The biphenyl dihydrodiol compound is then dehydrogenated to the dihydroxy compound by a dehydrogenase (*bphB*). The 2,3-dihydroxybiphenyl dioxygenase (*bphC*) catalyzes the ring-*meta*-cleavage at the 1,2 position to produce 2-hydroxy-6-oxo-6-phenylhexa 2,4-dienoic acid (HPDA). The yellow *meta*-cleavage compound is hydrolyzed to benzoic acid and 2-hydroxy-penta-2,4-dienoic acid by a hydrolase (*bphD*).

Genes involved in the metabolism of biphenyl (designated *bph*) was first cloned from *Pseudomonas pseudoalcaligenes* KF707 (Furukawa and Miyazaki, 1986) and thereafter from several other soil bacteria such as *P. paucimobilis* Q1 (Taira *et al.*, 1988), *Pseudomonas* sp. KKS102 (Kimbara *et al.*, 1989), *Pseudomonas* LB400 (Mondello, 1989), *Pseudomonas putida* OU83 (Khan and Walia, 1989), *P. putida* KF715 (Hayase *et al.*, 1990), *Pseutdomonas testosteroni* B356 (Ahmad *et al.*, 1990) and *Arthrobacter* sp. M5 (Peloquin and Greer, 1993). The 11.3 kb DNA fragment coding for the conversion of biphenyl to benzoic acid cloned from *P. pseudoalcaligenes* KF707 contains *bphA1A2A3A4*, *bphB*, *bphC* and *bphD* in this order (Taira *et al.*, 1992). In addition, there is a 3.5 kb-DNA segment (*bphX*) between *bphC* and *bphD* in the KF707 *bph* operon (Hayase *et al.*, 1990). Analyses of the *bphX* region reveal four genes which are involved in the metabolism of 2-hydroxy-penta-2,4-dienoate to acetyl Coenzyme A via a 4-hydroxy-2-oxo-valerate and acetaldehyde in the TCA cycle. Using the *bphABC* and *bphD* genes of *P. pseudoalcaligenes* KF707 as probes, Southern blot analyses revealed that almost identical *bph* genes are present in 7 strains among 15 strains tested (Furukawa *et al.*, 1989). The *bph* operons of *P. putida* KF715 and *Achromobacter xylosoxidans* KF701 are very similar to that of *P. pseudoalcaligenes* KF707, but the *bphX* (3.3 kb) region is absent between *bphC* and *bphD*. Some other biphenyl degrading strains possess weakly hybridized *bph* genes, but 3 strains did not show any homology. These results indicate that some chromosomal *bph* genes can be transferred among soil bacteria. Indeed, it was found that the *bph* operon of *P. putida* KF715 was transferred with high frequency to *P. putida* AC30 (Furukawa, unpublished data). It is also true that some *bph* genes accumulated mutations and show different degrees of homologies. Recently, it was demonstrated that

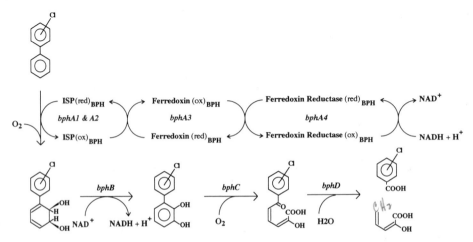

Figure 1. Metabolism of biphenyl by *P. pseudoalcaligenes* KF707 and *bph* genes involved in respective reactions.

some *bph* genes are greatly shuffled, as in the case of *Pseudomonas* sp. KKS102 *bph* operon (Fig. 2) (Kikuchi *et al.*, 1994). In the KKS *bph* operon, *bphA4* is located downstream of *bphD* and the *bphX* region (in KF707) is located upstream of *bphA1*.

COMPARISON OF THE *bph* OPERON AND TOLUENE CATABOLIC *tod* OPERON

P. putida F1 degrades toluene through *cis*-toluene dihydrodiol to 3-methylcatechol. The biochemistry and genetics of toluene metablism in this organism has been studied by Gibson and his associates (Zylstra et al. 1989). Nucleotide sequence determination revealed that *tod* operon of *P. putida* and *bph* operon of *P. pseudoalcalignes* KF707 are very similar in gene organization, as well as in size and homology of the corresponding enzymes, despite discrete substrate specificities of the two strains (Taira *et al.*, 1992). The amino acid sequences of the large and small subunits of the terminal dioxygenase, ferredoxin, ferredoxin reductase, dehydrogenase, and ring-*meta*-cleavage dioxygenase show ca. 60% identities between the two metabolic systems, but the similarity of the hydrolases (BphD *vs* TodF) is as low as 35% (Furukawa et al., 1993) (Fig. 3).

GENE COMPONENTS RESPONSIBLE FOR DISCRETE SUBSTRATE SPECIFICITY IN THE METABOLISM OF BIPHENYL AND TOLUENE

P. putida F1 grows well on toluene as a sole source of carbon and energy, but not on biphenyl. To the contrary, *P. pseudoalcaligenes* KF707 grows well on biphenyl, but not on

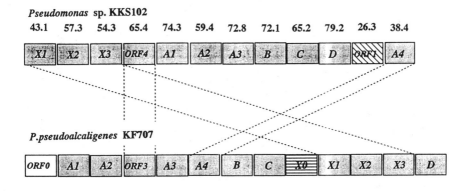

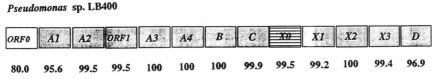

Figure 2. Comparison of *bph* operons among *P. pseudoalcaligenes* KF707, *Pseudomonas* sp. LB400, and *Pseudomonas* sp. KKS102. The numbers show amino acid identities (%) in comparison with the corresponding proteins of KF707.

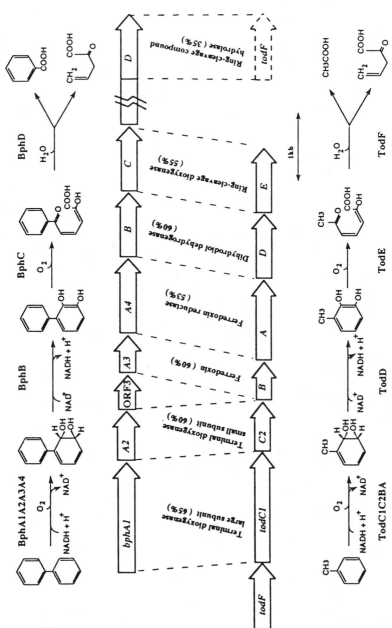

Figure 3. Comparison between biphenyl catabolic *bph* genes and toluene catabolic *tod* genes.

toluene. By using *E.coli* cells expressing *bphA1A2A3A4BC* genes, it was found that the inability of KF707 to utilize toluene was because that biphenyl dioxygenase cannot attack toluene. On the other hand, the inability of F1 to utilize biphenyl was because that hydrolase (TodF) is not capable of hydrolyzing the biphenyl *meta*-cleavage compound. In order to elucidate the critical components in the *bph*-encoded enzymes for toluene metabolism, a variety of *tod* genes were introduced into *P. pseudoalcaligenes* KF707. The introduction of *todC1C2* (the large and small subunits of toluene terminal dioxygenase gene, respectively) allowed the recombinant strain to grow on toluene, indicating that TodC1 and TodC2 forms a multicomponent dioxygenase associated with two other components of ferredoxin (BphA3) and ferredoxin reductase (BphA4) and this hybrid enzyme, composed of TodC1C2::BphA3A4, is functional for the initial dioxygenation of toluene to dihydrodiol. Toluene dihydrodiol can be converted to 3-methyl catechol by BphB (biphenyl dihydrodiol dehydrogenase). Further conversion of 3-methyl catechol through the *meta*-cleavage pathway is conducted by the benzene catabolic enzyme system (encoded by KF707 lower *bph* operon). The toluene utilizer *P. putida* F1 converted biphenyl into the yellow *meta*-cleavage compound (HPDA). Since the inability of F1 to grow on biphenyl is due to the lack of TodF activity for HPDA, introduction of *bphD* (KF707 HPDA gene) allowed F1 to grow on biphenyl (Furukawa *et al.*, 1993).

CONSTRUCTION OF HYBRID BIPHENYL (*bph*) AND TOLUENE (*tod*) GENES FOR FUNCTIONAL ANALYSIS OF AROMATIC RING DIOXYGENASES

Since we found that terminal dioxygenase (encoded by *bphA1A2* and *todC1C2*) and the ring-*meta*-cleavage compound hydrolase (encoded by *bphD* and *todF*) are critical for discrete substrate range between biphenyl-utilizing *P. putida* F1, we constructed a variety of hybrid gene clusters between *bph* and *tod* genes (Fig.4) to analyze the function of the hybrid aromatic ring dioxygenase in more detail (Hirose *et al.*, 1994). *E.coli* cells expressing the hybrid gene clusters, *tod::bphA2A3A4*, *todC1C2::bphA3A4* and *bphA1::todC2::bphA3A4*, gained the ability to convert benzene-toluene and their derivatives to the dihydrodiols, indicating that the hybrid terminal dioxygenase composed of TodC1::BphA2 and

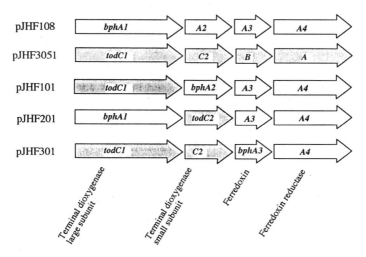

Figure 4. Schematic representation of *bph::tod* hybrid dioxygenase complexes. Open arrows show *bph* genes and shaded arrows show *tod* genes.

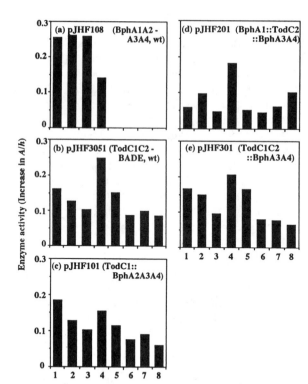

Figure 5. Production of dihydrodiol compounds from biphenyl, benzene, and their derivatives by *E. coli* cells carrying pJHF108 (a), pJHF101 (b), pJHF201 (c), pJHF301 (d), or pJHF3051 (e). Column 1, biphenyl (measured at 303 nm); column 2, 4-chlorobiphenyl (306 nm); column 3, 4-methylbiphenyl (307 nm); column 4, diphenylmethane (259 nm); column 5, naphthalene (262 nm); column 6, toluene (265 nm); column 7, 4-chlorotoluene (275 nm); column 8, benzene (260 nm).

BphA1::TodC2 forms a functionally active multicomponent dioxygnenase associated with ferredoxin (BphA3) and ferredoxin reductase (BphA4). Moreover, hybrid dioxygenase (composed of Tod C1::BphA2A3A4 and TodC1C2::BphA3A4) showed a wide substrate specificity rather similar to that of the wild-type toluene dioxygenase (TodC1C2BA) (Fig.5). On the other hand, the hybrid dioxygenase (BphA1::TodC2::BphA3A4) showed different oxidative activities for the same compounds, compared with other hybrid enzymes. These results suggest that (i) the two subunits of terminal dioxygenase are critically involved in the substrate specificity for biphenyl, benzene, and their derivatives, and (ii) the electron transport proteins, ferredoxin and ferredoxin reductase, are exchangeable with one another between biphenyl dioxygenase and toluene dioxygenase complexes.

CONCLUSION

It is postulated that many degraders of aromatic compounds could be involved in the final degradation of plant lignin, which is massively distributed in the environment and which consists of many polymerized aromatic moieties. This idea coincides with the fact that a number of catabolic genes involved in the degradation of aromatic compounds share a common ancestor and form gene super families (Harayama, 1989). The genetic diversity or shuffling of catabolic operons among soil bacteria is of particular interest from the viewpoint of how microorganisms gain novel catabolic activities for xenobiotics, which include many chemicals of man-made origin. Furthermore, the recent accumulation of sequence data will allow construction of hybrid enzymes which possess increased activity, or expanded substrate ranges. These approaches may provide an insight into the evolutionary relationship of catabolic functions in soil bacteria.

REFERENCES

Ahmad, D., Masse, R., and Sylvestre, M., 1990, Cloning and expression of genes involved in 4-chlorobiphenyl transformation by *Pseudomonas testosteroni*: homology to polychlorobiphenyl-degrading genes in other bacteria, *Gene 86*:53-61.

Furukawa, K., 1982, Microbial degradation of polychlorinated biphenyls. In: Chakrabarty AM (ed) *Biodegradation and Detoxification of Environmental Pollutants* pp. 33-57 CRC Press Inc., Boca Raton, Fla.

Furukawa, K., Hayase, N., Taira, K., and Tomizuka, N. 1989, Molecular relationship of chromosomal genes encoding biphenyl/polychlorinated biphenyl catabolism: some soil bacteria possess a highly conserved *bph* operon. *J. Bacteriol. 171*: 5467-5472.

Furukawa, K., Hirose, J., Suyama, A., Zaiki, T., and Hayashida, S. 1993, Gene components responsible for discrete substrate specificity in the metabolism of biphenyl (*bph* operon) and toluene (*tod* operon), *J. Bacteriol. 175*:5224-5232.

Harayama, S., and Kok, M. 1992, Functional and evolutionary relatioships among diverse organisms, *Annu. Rev. Microbiol. 46*:565- 601.

Hayase, N., Taira, K., and Furukawa, K., 1990, *Pseudomonas putida* KF715 *bphABCD* operon encoding biphenyl and polychlorinated biphenyl degradation: cloning, analysis and expression in soil bacteria, *J. Bacteriol. 172*: 1160-1164.

Hirose, J., Suyama, A., Hayashida, S., and Furukawa, K., 1994, Construction of hybrid biphenyl (*bph*) and toluene (*tod*) genes for functional analysis of aromatic ring dioxygenases, *Gene 138*: 27-33

Khan, A., and Walia, S. 1989, Cloning of bacterial genes specifying degradation of 4-chlorobiphenyl from *Pseudomonas putida* OU83, *Appl. Environ. Microbiol. 55*: 798-805

Kikuchi, Y., Yasukochi, Y., Nagata, Y., Fukuda, M., and Takagi, M., 1994, Nucleotide sequence and functional analysis of the *meta*-cleavage pathway involved in biphenyl and polychlorinated biphenyl degradation in *Pseudomonas* sp. strain KKS102, *J. Bacteriol., 176*:4269-4276.

Kimbara, K., Hashimoto, T., Fukuda, M., Koana, T., Takagi, M., Oishi, M., and Yano K (1989) Cloning and sequencing of two t andem genes involved in degradation of 2,3-dihydroxybiphenyl to benzoic acid in the polychlorinated biphenyl-degrading soil bacterium *Pseudomonas* sp.strain KKS102, *J. Bacteriol. 171*: 2740-2747

Mondello, F.J., 1989, Cloning and expression in *Escherichia coli* of *Pseudomonas* strain LB400 genes encoding polychlorinated biphenyl degradation, *J. Bacteriol. 171*: 1725-1732

Peloquinad, L., and Greer, C.W., 1993, Cloning and expression of the polychlorinated biphenyl-degradation gene cluster from *Ar/throbacter* M5 and comparison to analogous genes from Gram- negative bacteria, *Gene 125*: 35-40

Taira, K., Hayase, N., Arimura, N., Yamashita, S., Miyazaki, T., and Furukawa, K., 1988, Cloning and nucleotide sequence of the 2,3-dihydroxybiphenyl dioxygenase gene from the PCB-degrading strain of *Pseudomonas paucimobilis* Q1, *Biochemistry 27*:3990-3996

Taira, K., Hirose, J., Hayashida, S., and Furukawa, K., 1992, Analysis of *bph* operon from the polychlorinated biphenyl-degrading strain of *Pseudomonas pseudoalcaligenes* KF707, *J. Biol. Chem. 267*: 4844-4853

van der Meer, J.R., de Vos, W.M., Harayama, S., and Zehnder, A.J.B., 1992, Molecular mechanism of genetic adaptation to xenobiotic compounds, *Microbiol. Rev. 56*:677-694.

Zylstra, G.J., and Gibson, D.T., 1989, Toluene degradation by *Pseudomonas putida* F1: nucleotide sequence of the *todC1C2BADE* genes and their expression in *Escherichia coli. J. Biol. Chem. 264*: 14940- 14946

13

ADAPTATION OF SOYBEAN BRADYRHIZOBIA TO THE BRAZILIAN EDAPHIC SAVANNAHS

M. Cristina P. Neves,[1] Heitor L. C. Coutinho,[2] and Norma G. Rumjanek[1]

[1] EMBRAPA/National Center of Agrobiology Research (CNPAB)
P. O. Box 74.505, Seropédica, R J, Brazil
[2] Fundação Tropical de Pesquisa André Tosello
Campinas, S P, Brazil

INTRODUCTION

The vast Brazilian tropical edaphic savannah, referred to as Cerrado, is considered to be one of the last frontiers of agricultural utilization (Borloug and Dowswell, 1994). Soils are deep but leached of nutrients and, in general, strongly acidic (pH averages 5.0 and aluminum is present at toxic levels, Verdade, 1971). Agricultural utilization of the Cerrado was stimulated in the sixties when soil amendments, appropriated technologies, and improved/adapted varieties became available. To bring the land to crop production requires a ritual that includes land clearance (dragging or slash and burning of the native vegetation), ploughing, liming, and fertiliser application. In a little over two decades, the Cerrado became a major grain producing area, reducing social pressures threatening the very fragile soils of the Amazon region.

Nitrogen (N) fertilizer represents more than 50% of the cost of fertiliser for Brazilian farmers. The ability of legume plants, such as soybean (*Glycine max*), to establish a symbiotic relationship with diazotrophic bacteria known as bradyrhizobia, contributes to its self-sufficiency for N. Biological N fixation (BNF) can meet from 50 to 84% of the soybean N needs (Patterson and La Rue, 1983; Boddey et al., 1990). Amounts as high as 250 to 300 kg/ha N have been shown to be fixed by the soybean/bradyrhizobia symbiosis in the soils of the Cerrado (Boddey et al., 1990). Soybean has been considered an ideal crop for the Cerrado, and most of the success in the introduction of the new crop can certainly be attributed to the low initial fertiliser investment required.

COLONIZATION OF THE CERRADO WITH BRADYRHIZOBIA

The soybean root nodule bacterium has been classified as *Bradyrhizobium japonicum*. Recently a new species, *Bradyrhizobium elkanii,* was proposed (Kuykendall et al.,

Microbial Diversity in Time and Space, edited by Colwell et al.
Plenum Press, New York, 1996

1992) which characteristically produces rhizobitoxine that causes chlorosis in susceptible soybean cultivars (Devine et al., 1988) and has rhamnose and 4-O-methyl glucuronic acid as main components of the exopolysaccharide (EPS) (Minamisawa, 1989). The presence of the hydrogen-uptake system (Hup), enables recycling of the hydrogen evolved during BNF. It is considered advantageous to the symbiosis (Van Soon et al., 1993). Phenotype Hup$^+$ is mainly restricted to strains of *B. japonicum* (Minamisawa, 1989; Fuhrmann, 1990), but some exceptions have been reported (Basit et al., 1991). *B. elkanii* is widely distributed in soils. On the other hand, *B. japonicum* is more efficient and fixes more N (Fuhrmann, 1990)

It has been argued that soybean bradyrhizobia do not occur naturally in Brazilian soils (Vargas and Suhet, 1980). The first recorded introduction of these bacteria was in 1948 (Silva). Early trials, however, have consistently reported the superiority of local isolates over imported inoculants (Freire, 1982). Contaminated imported seeds may have played a role in spreading the bacteria into the fields (Vargas et al., 1993), although nodulation by native bradyrhizobia cannot be overruled.

Strains of several different origins were tested as inoculants for soybeans in the Cerrado. Most neither nodulated nor persisted in the trial sites. Nodulation failure was attributed to high host specificity of the cultivars and also to the low inoculant dose used, which was, however, the recommended dose for other regions (Vargas and Suhet, 1980).

A strain (BR-29; 29W) isolated at National Center of Agrobiology Research-EM-BRAPA was found to produce profuse nodulation in cultivars adapted to the Cerrado, when applied at high inoculum concentrations (5 times the usual concentrations). This strain may be a native bradyrhizobium (Neves, 1989). Strain BR-96 (SEMIA 587) was also successful, both in surviving and nodulating soybeans in the Cerrado. They comprised the commercial inoculant for soybeans in the Cerrado (Vargas and Suhet, 1980). Strain BR-40 (JF566) originally a component of the commercial inoculant, was introduced in the early trials. It failed to nodulate well and did not compete with other strains tested (Scotti et al., 1982). Strain BR-33 (CB1809), considered to be outstanding, in terms of BNF (Norris, 1967), also rated poorly in the early screening tests (Neves, 1989). It is a Hup$^+$ strain, producing nodules of high efficiency (N fixed per gram of nodule), enhancing the N harvest index of inoculated plants, reflected in yield increases of up to 30 above those obtained in plants inoculated with BR-29 (Neves et al., 1985). Strain BR-33 was found to promote a characteristic nodulation pattern in soybeans, defined as type 1, with the majority of nodules remaining pink and active well into the pod filling stage. Strains BR-29 and BR-96, on the other hand, showed a pattern defined as type 2, with a significant number of nodules presenting a central green spot, signalling senescence, just one month after sowing (Table 1).

The strains BR-29 and BR-96 have been characterised by 16S rRNA sequencing as *B. elkanii*, whereas BR-33 is a typical *B. japonicum* (see Table 1) (Rumjanek et al., 1993). Ability to multiply, compete with rhizosphere microflora, and survive saprophytically in the soil in the absence of the host plant are some characteristics determining the success of a strain colonising a new environment. Strains BR-29 and BR-96 successfully colonized the soils of the soybean growing area of Brazil. Their dominance in the soils represents an intriguing microbiological phenomenon and a cause of concern for scientists wishing to introduce other new isolates. Scotti et al. (1982) showed that successful strains (BR-29 and BR-96) are resistant to levels as high as 40 and 150 mg/l of streptomycin, respectively. On the other hand, the strains that failed to nodulate in the screening trials, BR-33 and BR-40, are sensitive to this antibiotic. The authors considered the possibility of the antibiotic resistance being a determinant in the survival and/or competitiveness of these strains in the Cerrado soils.

Table 1. Characteristics of selected *Bradyrhizobium* strains and isolates adapted to the Cerrado

Strain	Species	Hup phenotype	Serogroup	Nodulation pattern
BR-33	*B.japonicum*[a]		BR-33	Type 1c
BR-85	*B.japonicum*[b]		BR-33	Type 1
Isolate 70	*B.japonicum*[a]		BR-33	Type 1
BR-29	*B.elkanii*[a]	-	BR-29	Type 2
BR-96	*B.elkanii*[a]	-	BR-96	Type 2
BR-40	*B.elkanii*[b]	-	BR-40	Type 2
BR-86	*B.elkanii*[b]	-	BR-40	Type 2

[a] Rumjanek et al. 1993;

[b] Boddey and Hungria, 1994;

[c] Pereira, Rumjanek and Neves, unpublished results.

ADAPTATION OF STRAINS OF *BRADYRHIZOBIUM*

Chatel and Parker (1968) suggested that *Bradyrhizobium* strains would gradually adapt to infertile soils and harsh environments. In the Cerrado, increased dominance of one of the originally introduced strains (BR-40) has been observed. In 1970, this strain occupied approximately 2% of the nodules in the trial sites. Today, the dominance of strains BR-29 and BR-96 has been challenged by a sharp increase in the occurrence of BR-40, estimated in about 50% of the nodules (Vargas et al., 1993). This change in the bradyrhizobia population composition may reflect adaptation of strains to the environment, as well as changes in the soybean cultivars. The mechanisms involved in the process of survival and competition of the strains largely remain obscure and an improved understanding would help in devising strategies for establishment of improved strains. Screening isolates obtained in cultivated fields of the Cerrado, has led to the selection of naturally adapted isolates: isolate BR-85 (CPAC 7), belonging to serogroup BR-33 and isolates BR-86 (CPAC 15), 4A-5 and 4B-53, all belonging to serogroup BR-40 (Vargas et al., 1993; Scotti et al., 1993). Under controlled conditions, a significant increase, from 20 to 57% in the competitive ability of some adapted isolates, compared with the parental strain, has been reported (Scotti et al., 1993)

At the EMBRAPA-National Centre of Agrobiology Research, a strategy was devised to adapt strain BR-33 to the Cerrado (Neves et al., 1992). Sterile seeds of soybean, grown in limed and fertilised soil from recently cleared Cerrado and in soil containing either the naturalized strain BR 29 or BR 96. Nodules were selected by their relative efficiencies, i.e., values close to 1 (typical of strain BR-33) and serologically identified. The isolates were then used as inoculum for another planting cycle in new plots. The procedure was repeated several times.

The isolates maintained the characteristic efficiency of the original BR-33 strain (Neves et al.,1992), including the nodulation pattern described as type 1 (Table 1). Determination of the percentage of antagonistic actinomycetes present in the soil of the target area showed that most isolates were less affected by the actinomycetes in plate tests than the original strain, but no change in tolerance to aluminum or to low pH of the growth medium was observed (Neves et al.,1992). Abundant gum production is a distinguishing characteristic of strain BR-33, whereas strains BR-29 and BR-96 produce colonies of the large, watery type. Adapted isolates of BR-33 produced less gum, after exposure to the soil-plant system (Table 2; Neves et al., 1992).

The adapted isolates significantly improved soybean nodulation in the newly cleared soil of the Cerrado, compared to the parental strain (Neves et al., 1992), but only two strains

Table 2. Exopolysaccharide (EPS) and capsular polysaccharide (CPS) production by bradyrhizobia and discrimination of extracted polysaccharides using Pyrolysis Mass Spectrometry (Py-MS)

Strain/Isolate	EPS [α]		CPS [α]		Py-MS[β]
	ug/ml	ug/CFU.ml	ug/ml	ug/CFU.ml	(% similarly with V70)
BR33	33,0bc	0,96c	42,0a	1,22a	91
Isolate 72	37,5b	1,12a	36,3a	1,19ab	93
Isolate 68	34,2b	0,90c	17,4c	0,45cd	99
Isolate 70	60,7a	1,29b	28,0b	0,60c	-
BR33[strδ]	26,5cd	0,78d	37,2a	1,05b	ND[γ]
BR96	21,1d	0,56e	14,7c	0,39d	39

[α] Marinho et al., unpublished results; [β]Coutinho, 1993; [δ] spontaneous streptomycin resistant derivative; [γ]ND, not determined.

showed enhanced competitive ability and are included in a nationwide network trial of soybean strain selection that is in progress.

Actinomycetes are a major component of the microbial population in the Cerrado soil (Coelho and Drozdowicz, 1979). Disturbance of a natural soil ecosystem, such as when liming is applied, leads to changes in the microbial equilibrium favouring these organisms (Baldani et al. 1982; Döbereiner, 1982). This provides resistant microorganisms with an advantage in establishing themselves in the rhizosphere (Baldani et al., 1982; Scotti et al., 1982). Altered antibiotic resistance of some isolates was reported for concentrations of cells (Neves et al.,1992). Studies where inhibition by streptomycin was evaluated using single cell colonies have, however, shown that only one of the isolates was more resistant and able to grow in a medium containing sub-inhibitory concentrations of streptomycin (Coutinho, 1993). Furthermore, the existence of competitive isolates sensitive to the antibiotic suggests that changes in antibiotic resistance occurred during adaptation to the Cerrado may not be a pre-requisite for competitiveness in the Cerrado, leading to the conclusion that it may, in fact, be a result of changes in polysaccharide production, altering permeability of the cells to the drug, making it more resistant to the prevailing soil conditions.

Strains BR-29, BR-96, BR-40, BR-33 and their respective sub-strains were characterised using methods for analysis of the variation in both phenotype and genotype (Coutinho, 1993). DNA fingerprinting, using multiple arbitrary amplicon profiling (MAAP), as well as restricted fragment length polymorphism (RFLP), confirmed the origin of isolates serologically identified as a derivative of strain BR-33. These sub-strains, when analysed by pyrolysis mass spectrometry (Py-MS) fingerprinting, could be distinguished from the parental strain (Coutinho, 1993). Some of the derivatives that were serologically related to BR-40 and reported as having identical DNA fingerprinting (Coutinho, 1993), were found to differ from the parental strain in competitive ability (Scotti et al., 1993; Vargas et al., 1993), as well as in time of response to plant exudates, a necessary step in plant/bacteria dialogue leading to nodulation (Scotti et al., 1993). They also differed by demonstrating a greater intensity of one of the polypeptide bands in their membrane protein profiles (Scotti et al., 1993) and all were easily discriminated by Py-MS (Coutinho, 1993). Strain BR-33 was found to differ genetically from isolate BR-85 by one polymorphic restriction fragment, but belonged to the same serogroup.

The variability between strains and isolates by Py-MS of whole cell contents was also confirmed by analysis of polysaccharide extractions (see Table 2). Thus it was concluded that the important component of phenotypic variation, detected by Py-MS, arose from the difference in polysaccharide composition.

CONCLUSION

The strategy of speeding up natural adaptation has yielded an efficient strain, able to increase nodulation in a complex environment, subject to liming and fertiliser addition, where biotic and abiotic factors have limited strain introduction. The major characteristics distinguishing the adapted isolates from the parental strains were related to the bacterial cell membrane composition and to polysaccharide production, possibly altering permeability of the cells. Other characteristics, such as susceptibility to antibacterials produced by actinomycetes in the soil and antibiotic resistance are probably consequence of these alterations.

ACKNOWLEDGMENTS

CNPq scholarships to MCPN, NGR, HLCC and partial financial support from the Commission of the European Communities/ International Scientific Co-operation, Contract CI1*/0545, are gratefully acknowedged.

REFERENCES

Baldani, J. I., Baldani, V. L. D., Xavier, D. F., Boddey, R. M.., and Döbereiner, J., 1982, Efeito da calagem no número de actinomicetos e na porcentagem de bactérias resistentes a estreptomicina na rizosfera de milho, trigo e feijão, *Rev. Bras. Microbiol.*13: 250-263.

Basit, H.A., Angle, J.S., Salem, S., Gewaily, E.M., Kotob, S.I., and van Berkun, P., 1991, Phenotypic diversity among strains of *Bradyrhizobium japonicum* belonging to serogroup 110, *Appl. Environ. Microbiol.* 57:1570-1572.

Boddey, L.H., and Hungria, M., 1994, Classificação de estirpes de *Bradyrhizobium japonicum* em genótipo I e II baseada em características fenotípicas e genotípicas, in: *Proc. III Simpósio Brasileiro sobre Microbiologia do Solo*, p.66, IAPAR, Londrina,.

Boddey, R.M., Urquiaga, S., Suhet, A.R., Peres, J.R., and Neves, M.C.P. 1990, Quantification of the contribution of N_2 fixation to field-grown grain legumes: A strategy for the practical application of the ^{15}N isotope dilution technique, *Soil Biol. Biochem.* 22:649-655.

Borloug, N. E., and Dowswell, C.R., 1994, Feeding a human population that increasingly crowds a fragile planet, in: *Keynote Lecture, 15th World Congress of Soil Science*, International Society of Soil Science, Acapulco, Mexico.

Chatel, D. L., and Parker, C. A., 1968, Inhibition of rhizobia by toxic soil-water extracts, *Soil Biol. Biochem.* 4: 289-294.

Coelho, R.R.R., and Drozdowicz, A., 1979, The occurrence of actinomycetes in a cerrado soil in Brazil, *Rev. Ecol. Biol. Sol.* 15: 459-473.

Coutinho, H.L.C., 1993, *Studies of Bradyrhizobia from the Brazilian Cerrado,*. University of Bristol., PhD Thesis:

Devine, T.E., Kuykendall, L.D., and O'Neill, J.J.,1988, DNA Homology group and the identity of bradyrhizobial strains producing rhizobitoxine-induced foliar chlorosis on soybean, *Crop Sci.* 28:939-941.

Döbereiner, J., 1982, New evidence for the production and accumulation of antibiotics in nature, *Scripta Varia* 53.

Freire, J.R.J., 1982, Research into the Rhizobium/Leguminosae symbiosis in Latin America, *Plant Soil* 67:227-239.

Fuhrmann, J.,1990, Symbiotic effectiveness of indigenous soybean bradyrhizobia as related to serological, morphological, rhizobitoxine, and hydrogenase phenotypes, *App. Environ. Microbiol.* 56:224-229.

Kuykendall, L.D., Saxena, B., Devine, T.E., and Udell, S.E., 1992, Genetic diversity in *Bradyrhizobium japonicum* Jordan 1982 and a proposal for *Bradyrhizobium elkanii* sp. Nov., *Can.J.Microbiol.*38:501-505.

Minamisawa, K.,1989, Comparison of extracellular polysaccharide composition, rhizobitoxine production, and hydrogenase phenotype among various strains of *Bradyrhizobium japonicum*, *Plant Cell Physiol.* 30:877-884.

Neves, M.C.P., 1989, A new physiological approach for *Bradyrhizobium* strain selection, in: *Proc. World Soybean Research Conference IV*, vol. V, pp 2177-2192, Buenos Aires.

Neves, M.C.P., Didonet, A.D., Duque, F.F., and Döbereiner, J., 1985, *Rhizobium* strain effects on nitrogen transport and distribution in soybeans, *J. Exp. Bot.* 36: 1179-1192.

Neves, M.C.P., Ramos, M.L.G., Martinazzo, A.F., Botelho, G.R., and Döbereiner, J., 1992, Adaptation of more efficient soybean and cowpea rhizobia to replace established populations, in *Biological Nitrogen Fixation and Sustainability of Tropical Agriculture* (K. Mulongoy, M. Gueye, and D.S.C. Spencer, eds.), pp 219-233, Wiley-Sayce Co-Publ.

Norris, D.O., 1967, The intelligent use of inoculants and lime pelleting for tropical legumes, *Trop. Grassl.* 1:107-121.

Patterson, T.G., and La Rue, T.A., 1983, Nitrogen fixation by soybeans: Seasonal and cultivar effects, and comparison of estimates, *Crop Sci.* 23:488-492.

Rumjanek, N.G., Dobert, R.C., van Berkun, P., and Triplett, E.W., 1993, Common soybean inoculant strains in Brazil are members of *Bradyrhizobium elkanii, Appl. Environ. Microbiol.* 59: 4371-4373.

Scotti, M.R.M.M.L., Neves, M.C.P., Paiva, E., and Döbereiner, J., 1993, Effect of soybean roots on strain competitivity and protein profile of *Bradyrhizobium japonicum* adapted to Cerrado soils, *An. Acad. Bras. Ci.* 65:427-438.

Scotti, M.R.M.M.L., Sá, N.M.H., Vargas, M.A.T., and Döbereiner, J., 1982, Streptomycin resistance of *Rhizobium* isolates from Brazilian cerrados, *An. Acad. Bras. Cien.* 54:733-738.

Silva, J.G., 1948, Estudos sobre inoculação de soja, *Rev. Agric.* 22:365-378.

Van Soon, C. Rumjanek, N.G., Vanderleyden, J., and Neves, M.C.P., 1993, Hydrogenase in *Bradyrhizobium japonicum*: occurrence, genetics and effect on plant growth, *World J. Microbiol. Biotechnol.* 9:615-624.

Vargas, M.A.T.; Mendes, I.C.; Suhet, A.R., and Peres, J.R.R., 1993, Fixação biológica de nitrogênio, in: *A Cultura da Soja nos Cerrados,* (N.E. Arantes, and P.I.M. Souza, eds.), POTAFOS/EMBRAPA.

Vargas, M.A.T., and Suhet, S.R., 1980, Efeito de tipos e niveis de inoculantes na soja (*Glycine max*) cultivada em um solo de cerrados, *Pesq. Agropec. Bras.* 15:343-347.

Verdade, F.C., 1971, Agricultura e silvicultura no Cerrado, in: *III Simpósio sobre o Cerrado* (M.G. Ferri ed.), pp. 65-76, Ed. Universidade de São Paulo, São Paulo.

14

VIRUSES AND DNA IN MARINE ENVIRONMENTS

John H. Paul, Christina A. Kellogg, and Sunny C. Jiang

University of South Florida
Department of Marine Science
140 Seventh Avenue South
St. Petersburg, Florida 33701

INTRODUCTION

DNA is the genetic material found in all living cells and is a molecule fundamental for life on this planet. Interest in the measurement of DNA in the oceans has evolved from a desire to quantitate biomass to a need to understand microbial diversity in the marine environment (see chapter by DeLong in this book).

The first measurements of DNA in the water column were directed at estimating phytoplankton biomass (Holm-Hansen et al., 1968; Holm-Hansen, 1969). Subsequent work employing size fractionation and measurement of DNA, chlorophyll a, and bacterial direct counts indicated that most of the DNA was associated with heterotrophic bacterioplankton rather than phytoplankton (Paul and Carlson, 1984; Paul et al., 1985).

Today we understand that DNA in the marine environment exists in two operationally defined forms—particulate and dissolved, depending on whether the material passes through a 0.2 μm filter (Jiang and Paul, 1995; Figure 1). The particulate fraction is associated primarily with small microbial cells (heterotrophic bacterioplankton and autotrophic pico-plankton), with 70-99% being < 1 μm in size. The composition of dissolved DNA is not quite as straightforward.

DISSOLVED DNA IN MARINE ENVIRONMENTS

Dissolved DNA is a ubiquitous component of the dissolved organic matter (DOM) in aquatic environments, generally found at concentrations of 5-44 μg/l for estuarine environments, 1-5 μg/l for open ocean waters, and < 1 μg/l for deep sea environments (Pillai and Ganguly, 1970; Deflaun et al., 1987; Karl and Bailiff, 1989). The molecular weight range was found to be < 500 bp to > 23 kbp (Deflaun et al., 1987). However, these studies predated the understanding of the abundance of viral particles in aquatic environments (Bergh et al., 1989; Proctor and Fuhrman, 1990). When making simultaneous measurements of dissolved

Microbial Diversity in Time and Space, edited by Colwell et al.
Plenum Press, New York, 1996

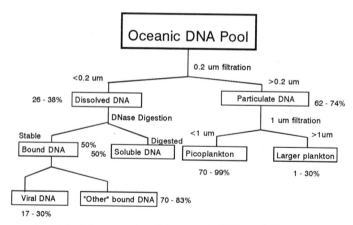

Figure 1. Conceptual model of the properties of the oceanic DNA pool. Percentages are additive at each horizontal level. (Jiang and Paul, 1995).

DNA and viral abundance, our preliminary studies (Paul *et al.*, 1991) suggested that viruses (or DNA encapsulated in viral particles) was a relatively small component of the dissolved DNA. However, using a highly sensitive direct measure of dissolved DNA, Maruyama *et al.* (1993) found most of the DNA to be DNase insensitive (termed "coated"), which they attributed to viral DNA. Using an ultracentrifugation technique to separate viral DNA (vDNA) from soluble DNA (sDNA), Beebee *et al.* (1991) concluded that most (>80%) of the filterable DNA was in the vDNA fraction. Neither of these groups enumerated viral particles in their studies, to test the validity of the conclusion that "coated DNA" was in fact vDNA. Using a combination of DNase digestion, differential centrifugation, and ethanol precipitation to separate various components, we have concluded that dissolved DNA appears to be comprised of at least three forms: 1) free or naked DNA (sDNA), making up at least half of the dissolved DNA; this form of DNA, which turns over rapidly, is thought to be the result of cell lysis, from senescence or viral infection and excretion from viable cells (Hara and Ueda, 1981); 2) DNA encapsulated in viral particles (vDNA), which comprises < 1-22% of the dissolved DNA, and 3) an uncharacterized "bound" form, making up the remainder. This bound form may be DNA in colloids (Wells and Goldberg, 1991, 1993) or membrane "blebs" (Dorwood *et al.*, 1989; Dorwood and Garon, 1989; 1990; Figure 1).

Besides merely quantitating dissolved DNA and determining its origin, we have been interested in the genetic consequences of dissolved DNA as a source of transforming DNA for gene transfer and the evolution of microbial communities (Paul *et al.*, 1992; Frischer *et al.*, 1994) and, more recently, in the genetic significance of viruses in the oceans. It is important to understand genetic diversity of the simplest form of life in the oceans, namely the viruses. Viruses can also exert a genetic influence on their host (and vice versa) by the process of lysogeny.

GENETIC DIVERSITY OF VIBRIOPHAGES

The advent of DNA technology has added tools such as DNA-DNA homology, restriction fragment patterns, and genomic DNA molecular weights (Werquin, *et al.*, 1988; Lindstrom and Kaijalainen, 1991) to the traditional diagnostics of size, morphology, serology, and physiological properties (Adams, 1952), commonly used to assess the diversity of viruses. A combination of these methods, used to analyze eucaryotic algal viruses, led to the discovery of groups of viruses from diverse locations, which share a common host and are similar or identical in morphology, but have considerable variability within their genomes (Schuster, *et al.*, 1986; Cottrell and Suttle, 1991). Similar analysis was performed using five cyanophages which infect *Synechococcus* (Wilson *et al.*, 1993). Limited homology under low stringency probing conditions revealed a low level of relatedness among the five viruses isolated from the Sargasso Sea, Woods Hole Harbor, MA, and the English Channel.

Other than the studies mentioned above, there has not been much research in the area of phylogeography of marine viruses, particularly phages. This is especially surprising in the case of vibriophages, since this family (*Vibrionaceae*) contains the greatest number of reported phage-host systems for the marine environment (Moebus, 1987), with the genus *Vibrio* comprising most of the hosts (Moebus and Nattkemper, 1983). Since the first isolation of a bacteriophage specific for *Vibrio parahaemolyticus* (Nakanishi *et al.*, 1966), these phage-host systems have received much attention (Baross and Liston, 1968; Kaneko and Colwell, 1973; Sklarow *et al.*, 1973; Baross *et al.*, 1978a; Baross *et al.*, 1978b; Hidaka and Tokushige, 1978; Koga and Kawata, 1981; Koga *et al.*, 1982). These studies, however, were largely concerned with the morphology, sensitivity to heat and chemicals, burst size, and host range of such phages. We chose instead to conduct a systematic study with a large sample size, to determine the geographic distribution and genetic diversity of marine vibriophages that infect a host *Vibrio parahaemolyticus* (strain 16) isolated from Tampa Bay, FL.

Over 70 lytic phages have been isolated using Host 16 from various marine environments, covering a geographic distance of more than 4,500 miles (Table 1). In addition to sharing a common host and being morphologically similar, these viruses appear to be genetically related, based on stringent hybridization of their DNA with a 1.5 kb EcoR1 fragment of the Φ16 (the type-phage from Tampa Bay, FL) genome, used as a gene probe. The nature of this fragment is cryptic, yet it is shared by all isolates characterized to date. When the various phage genomes were digested with EcoRI, and the digests probed with the 1.5 kb fragment, several banding patterns were observed. Phages with common banding patterns were segregated into groups. The phage groups are not identical, evidenced by their differing restriction patterns and autoradiography patterns (Figure 2), yet clearly share sequence homology of at least 1.5 kb [≥ 2.6% of the genome]. The 1.5 kb gene probe has not been found to hybridize with DNA from other phages (namely T2 and several environmental isolates), ruling out a "trans-species" function, such as DNA polymerase. It also does not hybridize with DNA from marine vibriophages specific to *V. parahaemolyticus* strains HER1165 and HER1169 (Vp1, Vp5, Vp6, Vp11, Vp12), suggesting that this fragment may be unique to the Φ16-like group of vibriophages.

The rationale of screening viral isolates by probing EcoRI digests with this 1.5 kb fragment was to sort the phages rapidly into groups (Table 2, Figure 2). The 1.5 kb common region from a representative of each group could then be PCR amplified, cloned, and sequenced. A comparison of these sequences would allow an estimate of genetic divergence between the "genotypes" to be calculated from the similarities of their DNA sequences.

In light of evidence that phage-host systems can reflect different water masses (Hidaka and Sakita, 1981; Moebus and Nattkemper, 1983; Moebus, 1983), we have speculated that the distribution of these phages in the Gulf of Mexico and around the Florida peninsula may be related to the path of the Florida Current, which curves around the state before heading north to form the Gulf Stream. While there was clearly no sequential geographic distribution of the restriction banding pattern groups, advection by the Florida Current could explain how Group E phages can be found in Key Largo, Tampa Bay, offshore in the Gulf of Mexico, and in the Fort Jefferson moat in the Dry Tortugas.

Table 1. Vibriophage isolation locations and dates (Kellogg *et al.*, 1995)

Phages	Isolated from	Date
Φ16	Tampa Bay, St. Pete. Pier, FL	1991
ΦKL3, ΦKL5	Canal, Key Largo, FL	January, 1992
ΦKL6, ΦKL7		
ΦKL33, ΦKL34	Blackwater Sound, Key Largo, FL	January, 1992
ΦKL35, ΦKL36		
ΦKL44	Tarpon Sound, Key Largo, FL	January, 1992
ΦPEL1A-1, ΦPEL1A-2	Mouth of Tampa Bay, FL	July, 1992
ΦPEL8A-1, ΦΦPEL8A-2	Gulf of Mexico, St. 8, Surface waters	July, 1992
ΦPEL8A-3		
ΦPEL8C-1, ΦPEL8C-2	Gulf of Mexico, St. 8, 1500 meters	July, 1992
ΦPEL13A-1	Off Garden Cay, Dry Tortugas	July, 1992
ΦPEL13B-1, ΦPEL13B-2	Moat, Ft. Jefferson, Dry Tortugas	July, 1992
ΦPEL13B-3, ΦPEL13B-4		
ΦPEL13B-5, ΦPEL13B-6		
ΦPEL13B-7, ΦPEL13B-8		
ΦPEL13B-9, ΦPEL13B-10		
ΦMoat-1	Moat, Ft. Jefferson, Dry Tortugas	June, 1993
ΦMarq-1, ΦMarq-2	Marquesas	June, 1993
ΦMarq-3		
ΦKWH-2, ΦKWH-3	Key West Harbor, FL	June, 1993
ΦKWH-4		
ΦHAWI-1, ΦHAWI-2	Ala Wai Canal, Honolulu, HI	October, 1993
ΦHAWI-3, ΦHAWI-4		
ΦHAWI-5, ΦHAWI-6		
ΦHAWI-7, ΦHAWI-8		
ΦHAWI-9, ΦHAWI-10		
ΦHD0-1, ΦHD0-2	Ke'ehi Lagoon, HI	October, 1993
ΦHD0-3, ΦHD0-4	Ke'ehi Lagoon, HI	February, 1994
ΦHD0-5, ΦHD0-6		
ΦHD1S-1, ΦHD1S-2	Sand Island Transect, HI	February, 1994
ΦHD2S-1, ΦHD2S-2	Sand Island Outfall, HI	February, 1994
ΦHD2S-3, ΦHD2S-4		
ΦHD2S-5		
ΦHC1-1, ΦHC1-2	Upper Pearl Harbor, HI	February, 1994
ΦHC1-3, ΦHC1-4		
ΦHC2-1, ΦHC2-2	Mid Pearl Harbor, HI	February, 1994
ΦHC2-3, ΦHC2-4		
ΦHC3-1, ΦHC3-2	Mouth of Pearl Harbor, HI	February, 1994
ΦHC3-3		

A mechanism of long-distance transport, other than natural currents exists in the form of cargo ships and freighters which cross the Panama Canal. Such vessels often take on seawater as ballast, and the local microbial and planktonic population are unnoticed passengers (Hallegraeff and Bolch, 1992; McCarthy and Khambaty, 1994). Mamala Bay, HI and Tampa Bay, FL are both ports of call for these types of ships; and ships en-route to Miami or further east would have to go around the Dry Tortugas, Marquesas, and Florida Keys on their way out of the Gulf of Mexico. Interestingly, these phages were isolated only from harbor waters (Pearl Harbor, Ke'ehi Lagoon near Honolulu Harbor) in Hawaii. No Φ16-like vibriophages were isolated from other coastal environments in Hawaii.

Hidaka (1980) found a *Vibrio* phage-host system with a fairly wide oceanic distribution: the host was isolated near Hawaii, but the phage was found in waters just south of Japan. This study corroborates that such "long-distance" phage-host systems may not be uncommon. Results of our research confirm that there are identifiable genetic populations of the

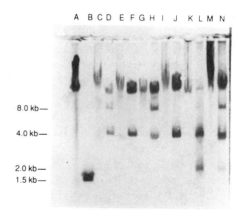

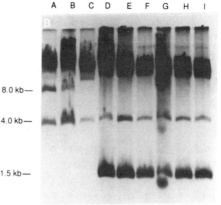

Figure 2. Autoradiograms of EcoRI digested phage DNA which have been probed with the 1.5 kb fragment. (a) A = uncut Φ16; B = digested Φ16; C = uncut ΦPEL1A-1; D = digested ΦPEL1A-1; E = uncut ΦPEL1A-2; F = digested ΦPEL1A-2; G = uncut ΦPEL8A-1; H = digested ΦPEL8A-1; I = uncut ΦPEL8A-2; J = digested ΦPEL8A-2; K = uncut ΦPEL8A-3; L = digested ΦPEL8A-3; M = uncut ΦPEL8C-1; N = digested ΦPEL8C-1. (b) All EcoRI digests, no uncut DNA's: A = ΦKL7; B = ΦPEL8C-1; C = ΦPEL13B-1; D = ΦHAWI-10; E = ΦHD0-1; F = ΦHD1S-1; G = ΦHD2S-1; H = ΦHC1-2; I = ΦHC2-3. Note: The 2.0 kb band of ΦPEL8C-1 is not visible in this film due to incomplete digestion of that sample. (Kellogg *et al.*, 1995).

Table 2. Groupings of phage isolates based upon similar EcoR1 restriction banding patterns detected by autoradiography after probing with the Φ16 gene probe (Kellogg *et al.*, 1995)

Super group	Phages	Visible bands[1]	Molecular weights (KB)
A	Φ16	1	1.5
B	34 Hawaii phages, ΦKL33	2	4.0, 1.5
C	ΦKL3, ΦKL5, ΦKL44, ΦKL7, ΦKL34, ΦKL6, ΦPEL 13A-1, ΦPEL 1A-1, ΦPEL 8C-1, ΦPEL 8A-1, ΦMARQ-1, ΦKWH-2, ΦKWH-3, ΦKWH-4	2	8.0, 4.0
D	ΦPEL 8A-3, ΦPEL 8C-1	3	8.0, 4.0, 2.0
E	ΦKL36, ΦPEL 1A-2, ΦPEL 8A-2, ΦPEL 13B-1 TO 5,ΦPEL 13B-7 TO 10, ΦMOAT-1	1	4.0
F	ΦKL35, ΦPEL 13B-6	1	4.2

[1] Excluding high molecular weight (>23 kb) which may be undigested genome.

Φ16-type vibriophages not only in the Gulf of Mexico, Tampa Bay, and Florida Keys, but also in coastal waters of Hawaii, *ca.* 4,500 miles distant. Additionally, we have found that these populations are closely related (62-80% nucleotide similarity of a 340 bp sequence), suggesting that virus "species", such as the vibriophages, may be widely distributed in the global ocean.

LYSOGENY IN THE MARINE ENVIRONMENT

Lysogeny occurs when a temperate phage enters into a stable symbiosis with its host bacterium, integrating into the host chromosome as a prophage, rather than beginning a cycle of cell lysis. Presence of the temperate phage genome prevents development of further lytic infections by the same phage, through production of repressors, termed *immunity to super-infection*.

A review of the literature by Ackermann and DuBow (1987) indicates that, of nearly 1200 strains of bacteria investigated, an average of 47% contained inducible prophage (were lysogenized). A survey of *Pseudomonas* cells isolated from a lake showed that 70% of the isolates contained DNA homologous to the temperate phage F116, suggesting significant occurrence of lysogens (Ogunseitan *et al.*, 1990). All groups of bacteria have been reported to contain lysogenic members, including archaebacteria, and cyanobacteria, as well as some protozoans (Preer *et al.*, 1971). Polylysogeny, or the ability to contain more than one type of temperate phage, is also common (Bertani, 1953). Where examined, 2-68% of the strains were polylysogenic, with some strains of *Bacillus megaterium* containing five different phages (Ackermann and DuBow, 1987). Lytic inductions, which result in defective phage particles (empty heads, tails only, and heads only; see Figure 3B) are believed to be the result of *bacteriocinogeny*, and the particles produced are termed *bacteriocins* (Kageyama, 1975).

The lysogenic state is an highly evolved state or condition (Stewart and Levin, 1984). The process of prophage replication and superinfection immunity require coordinate expression of prophage and host genes (Stewart and Levin, 1984), with substantial evidence that temperate phages and their hosts have co-evolved (Levin and Lenski, 1983).

Lysogeny has been found to be common in every environment thus far examined. To determine the prevalence of lysogeny in the marine environment, we screened 88 randomly isolated bacterial strains from Tampa Bay, the southeastern Gulf of Mexico, Key Largo, the Dry Tortugas, and most recently Mamala Bay, Oahu, for the presence of inducible prophage, using mitomycin C induction (Jiang and Paul, 1994). The results of this study are summarized in Table 3. The percentage of inducible lysogens amongst bacterial isolates ranged from 25% to over 63%, averaging 42%. This is in agreement with the percentage of lysogens amongst 1200 strains cited by Ackermann and DuBow (1987). In general, the lower percentages were found in eutrophic waters (Tampa Bay, Florida Keys, Sewage outfall), and relatively higher numbers in oligotrophic offshore environments. This distribution pattern substantiates laboratory results (Borek, 1952; Edlin *et al.*, 1977; Schrader *et al.*, 1994) suggesting low nutrient conditions favor lysogeny or that lysogens outcompete non-lysogens in nutrient-limited environments. We have also found that continuous subculture of the isolates in nutrient rich medium may cause loss of inducibility, in some instances.

Transmission electron microscopy (TEM) examination of the induced bacterial lysates revealed a variety of particle morphologies, with two major groups of phage-like particles observed. One type consisted of complete phage particles with electron-dense heads (presumably containing nucleic acid material), tail, and tail fibers, as found in the induction lysate of an oligotrophic Gulf of Mexico isolate (Figure 3A); the other type was characterized by broken phage tails, empty phage heads, or both, as seen in Figure 3B, which came from

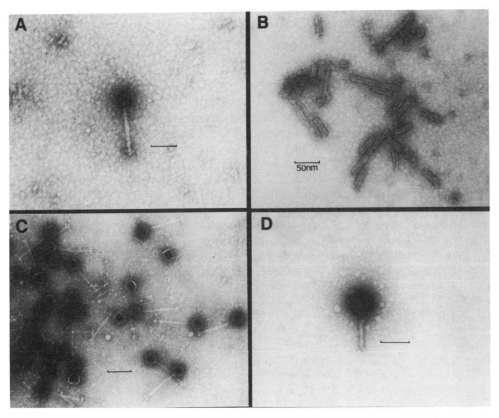

Figure 3. Particles observed in the supernatants of mitomycin C induced bacterial lysates (Panels A and B) and temperate phages isolated from marine environments (Panels C and D). The lysogenic bacterium induced in panel A was isolated from the Gulf of Mexico. The other phages and hosts (Panels B to D) were isolated from Mamala Bay, Oahu, Hawaii. The bar represents 100 nm in pictures unless indicated in the picture.

Table 3. Occurrence of lysogeny/bacteriocinogeny (Lys./Bac.) amongst bacterial isolates as determined by mitomycin C induction

Sampling sites	No. of Isol. examined	No. of Lys./Bac.	% (Lys./Bac.) /Isol.
Florida, USA			
Tampa Bay, Estuarine	23	6	26.1%
Florida Keys,			25.0%
Coastal Eutrophic Zone	12	3	
Florida Keys, Oligotrophic			63.6%
and Reef Environments	22	14	
Oligotrophic Southeastern			42.9%
Gulf of Mexico	7	3	
Mamala Bay, Hawaii			
Sand Island Sewage Outfall	10	3	30.0%
Pearl Harbor	6	3	50.0%
Offshore Diamond Head	8	5	62.5%
Total	88	37	42.0%

the lysate of a bacterial isolate from Mamala Bay, Oahu, Hawaii, believed to be are defective phage particles or bacteriocins.

SUMMARY AND CONCLUSIONS

The role of viruses in the marine environment and their interactions with DNA found in the ocean is just beginning to be elucidated. The question of biodiversity of the simplest form of life in the oceans is unfolding. We present evidence here for the existence of very closely related populations of vibriophages separated by distances in excess of 4500 miles and argues for the existence of circumglobal phage species in the oceans. Lysogeny is also apparently widespread in the marine environment, and nearly half of the marine bacteria may harbor a silent viral infection in the form of a prophage. The effect of lysogeny on the fitness of marine bacteria and relationship to starvation survival and competition with non-lysogenized counterparts are completely unknown. These areas will be fruitful fields for investigation in the future and should contribute to better understanding of the biodiversity of microorganisms in time and space.

REFERENCES

Ackermann, H. W-., and DuBow, M. S., 1987, *Viruses of Prokaryotes, Vol 1. General Properties of Bacteriophages*, CRC Press, Boca Raton.

Adams, M. H., 1952, Classification of bacterial viruses: characteristics of the T5 species and of the T2, C16 species, *J. Bact.* 64: 387-396.

Baross, J. A., and Liston, J., 1968, Isolation of *Vibrio parhaemolyticus* from the Northwest Pacific, *Nature 217*: 1263-1264.

Baross, J. A., Liston, J., and Morita, R. Y., 1978a, Incidence of *Vibrio parahaemolyticus* bacteriophages and other *Vibrio* bacteriophages in marine samples, *Appl. Environ. Microbiol.* 36: 492-499.

Baross, J. A., Liston, J., and Morita, R. Y., 1978b, Ecological relationship between *Vibrio parahaemolyticus* and agar-digesting vibrios as evidenced by bacteriophage susceptibility patterns, *Appl. Environ. Microbiol.* 36: 500-505.

Beebee, T. J. C., 1991, Analysis, purification and quantification of extracellular DNA from aquatic environments, *Fresh. Biol.* 25: 525-532.

Bergh, O., Borsheim, K. Y., Bratbak, G., and Heldal, M., 1989, High abundance of viruses found in aquatic environments, *Nature 340*: 467-468.

Bertani, G., 1953, Lysogenic versus lytic cycle of phage multiplication, *Cold Spring Harb. Symp. Quant. Biol.* 18: 65-70.

Borek, E., 1952, Factors controlling aptitude and phage development in a lysogenic *Escherichia coli* K12, *Bichim. Biophys. Acta 8*: 211-215. 184.

Cottrell, M. T., and Suttle, C. A., 1991, Wide-spread occurrence and clonal variation in viruses which cause lysis of a cosmopolitan eukaryotic marine phytoplankter, *Micromonas pusilla, Mar. Ecol. Prog. Ser.* 78: 1-9.

DeFlaun, M. F., Paul, J. H., and Jeffrey, W. H., 1987, The distribution and molecular weight of dissolved DNA in subtropical estuarine and oceanic environments, *Mar. Ecol. Prog. Ser. 33*: 29-40.

Dorwood, S. W., Garon, C. F., and Judd, R. C., 1989, Export and intercellular transfer of DNA via membrane blebs of *Neisseria gonorrhoeae, J. Bact. 171*: 2499-2505.

Dorwood, S. W., and Garon, C. F., 1989, DNA-binding proteins in cells and membrane blebs of *Neisseria gonorrhoeae, J. Bact. 171*: 4196-4201.

Dorwood, S. W., and Garon, C. F., 1990, DNA is packaged within membrane-derived vesicles of Gram-negative but not Gram-positive bacteria, *Appl. Environ. Microbiol.* 56: 1960-1962.

Frischer, M. E., Stewart, G. J., and Paul, J. H., 1994, Plasmid transfer to indigenous marine bacterial populations by natural transformation, *FEMS Microb. Ecol.* 15: 127-136.

Hallegraeff, G. M., and Bolch, C. J., 1992, Transport of diatom and dinoflagellate resting spores in ships' ballast water: implications for plankton biogeography and aquaculture, *J. Plank. Res. 14*: 1067-1084.

Hara, T., and Ueda, S., 1981, A study on the mechanism of DNA excretion by *Pseudomona aeruginosa* KYO-1: Effect of mitomycin C on extracellular DNA production, *J. Agric. Biol. Chem. 45*: 2457-2461.

Hidaka, T., 1980, Analytical research of microbial ecosystems in seawater around fishing ground-II. On a residentary bacteriophage-system in seawater around the Ryuku Island Arc, *Mem. Fac. Fish., Kagoshima Univ. 29*: 327-337.

Hidaka, T., and Sakita, I., 1981, Analytical research of micriobial ecosystems in seawater around fishing ground-II. On the habitat segregation of bacteriophage systems in the west region of the northern Ryuku Island Arch, *Mem. Fac. Fish. Kagoshima Univ. 30*: 331-338.

Hidaka, T., and Tokushige, A., 1978, Isolation and characterization of *Vibrio parahaemolyticus* bacteriophages in seawater, *Mem. Fac. Fish., Kagoshima Univ. 27*: 79-90.

Holm-Hansen, O., Sutcliffe, W. H., and Sharp, J., 1968, Measurement of deoxyribonucleic acid in the ocean and its ecological significance, *Limnol. Oceanogr. 13*: 507-514.

Holm-Hansen, O., 1969, Determination of microbial biomass in ocean profiles, *Limnol. Oceanogr. 14*: 740-747.

Jiang, S. C., and Paul, J. H., 1994, Seasonal and diel abundance of viruses and occurrence of lysogeny/bacteriocinogeny in the marine environment, *Mar. Ecol. Prog. Ser. 104*: 163-172.

Jiang, S., and Paul, J. H., 1995, Viral contribution to dissolved DNA in the marine environment: differential centrifugation and kingdom probing, *Appl. Environ. Microbiol. 61*: 317-325.

Kageyama, M., 1975, Bacteriocins and bacteriophages in *Pseudomonas aeruginosa*, In: Mitsuhashi, S., Hashimoto, H. (eds.) *Microbial Drug Resistance*, University of Tokyo Press, Tokyo, p. 291-305.

Kaneko, T., and Colwell, R. R., 1973, Ecology of *Vibrio parahaemolyticus* in the Chesapeake Bay, *J. Bact. 113*: 24-32.

Karl, D. M., and Bailiff, M. D., 1989, The measurement and distribution of dissolved nucleic acids in aquatic environments, *Limnol. Oceanogr. 34*: 543-558.

Kellogg, C. A., Rose, J. B., Jiang, S. C., Thurmond, J. M., and Paul, J. H., 1995, Genetic diversity of related vibriophages isolated from marine environments around Florida and Hawaii, USA, *Mar. Ecol. Prog. Ser. 120*: 89-98.

Koga, T., and Kawata, T., 1981, Structure of a novel bacteriophage VP3 for *Vibrio parahaemolyticus*, *Microbiol. Immunol. 25*: 737-740.

Koga, T., Toyoshima, S., and Kawata, T., 1982, Morphological varieties and host ranges of *Vibrio parahaemolyticus* bacteriophages isolated from seawater, *Appl. Environ. Microbiol. 44*: 466-470.

Levin, B. R., and Lenski, R. E., 1983, Coevolution in bacteria and their viruses and plasmids, In: Futuyama, D. J., Slaktin, M. (eds.) *Coevolution*, Sinauer and Associates, Inc., Sunderland, Massachusetts, p. 99-127.

Lindstom, K., and Kaijalainen, S., 1991, Genetic relatedness of bacteriophage infecting *Rhizobium galegae* strains, *FEMS Microbiol. Let. 82*: 241-246.

Maruyama, A., Oda, M., and Higashihara, T., 1993, Abundance of virus-sized non- DNase-digestible DNA (coated DNA) in eutrophic seawater, *Appl. Environ. Microbiol. 59*: 712-717.

McCarthy, S. A., and Khambaty, F. M., 1994, International dissemination of epidemic *Vibrio cholerae* by cargo ship ballast and other nonpotable waters, *Appl. Environ. Microbiol. 60*: 2597-2601.

Moebus, K., 1983, Lytic inhibition responses to bacteriophages among marine bacteria, with special reference to the origin of phage-host systems, *Helgolander Meeresunters. 36*: 375-391.

Moebus, K., 1987, Ecology of marine bacteriophages, In: Goyal, S. M., Gerba, C. P., Bitton, G. (eds.) *Phage Ecology*, John Wiley & Sons, New York, p. 137-156.

Moebus, K., and Nattkemper, H., 1983, Taxonomic investigations of bacteriophage sensitive bacteria isolated from marine waters, *Helgolander Meeresunters. 36*: 357- 373.

Nakanishi, H., Iida, Y., Maeshima, K., Teramoto, T., Hosaka, Y., and Ozaki, M., 1966, Isolation and properties of bacteriophages of *Vibrio parahaemolyticus*, Biken. *J. 9*: 149-157.

Ogunseitan, O. A., Sayler, G. S., and Miller, R. V., 1990, Dynamic interaction of *Pseudomonas aeruginosa* and bacteriophages in lake water, *Microb. Ecol. 19*: 177- 185.

Paul, J. H., and Carlson, D. J., 1984, Genetic material in the marine environment: implication for bacterial DNA, *Limnol. Oceanogr. 29*: 1091-1097.

Paul, J. H., Jeffrey, W. H., and DeFlaun, M. F., 1985, Particulate DNA in subtropical oceanic and estuarine planktonic environments, *Mar. Biol. 90*: 95-101.

Paul, J. H., Jiang, S. C., and Rose, J. B., 1991, Concentration of viruses and dissolved DNA from aquatic environments by vortex flow filtration, *Appl. Environ. Microbiol. 57*: 2197-2204.

Paul, J. H., Thurmond, J. M., Frischer, M. E., and Cannon, J. P., 1992, Intergeneric natural plasmid transformation between *E. coli* and a marine *Vibrio* species, *Molec. Ecol. 1*: 37-46.

Pillai, T. N. V., and Ganguly, A. K., 1970, Nucleic acids in the dissolved constituents of sea-water, *Curr. Sci. (India) 22*: 501-504.

Preer, J. R., Preer, L. B., Rudman, B., and Jurand, A., 1971, Isolation and composition of bacteriophage-like particles from kappa of killer paramecia, *Mol. Gen. Genet. 111*: 202.

Proctor, L. M., and Fuhrman, J. A., 1990, Viral mortality of marine bacteria and cyanobacteria, *Nature 343*: 60-62.

Schrader, J. O., Lufburrow, M. D., and Kokjohn, T. A., 1994, Effects of stress on the replication potential of bacteriophages, *94th General Meeting of the American Society for Microbiology*, Las Vegas, Abs. N-213.

Schuster, A. M., Burbank, D. E., Meister, B., Skrdla, M. P., Meints, R. H., Hattman, S., Swinton, D., and Van Etten, J. L., 1986, Characterization of viruses infecting a eukaryotic *Chlorella*-like green alga, *Virol. 150*: 170-177.

Sklarow, S. S., Colwell, R. R., Chapman, G. B., and Zane, S. F., 1973, Characteristics of a *Vibrio parahaemolyticus* bacteriophage isolated from Atlantic coast sediment, *Can. J. Microbiol. 19*: 1519-1520.

Stewart, F. M., and Levin, B. R., 1984, The population biology of bacterial viruses: why be temperate, *Theor. Popul. Biol. 26*: 93-117.

Wells, M. L., and Goldberg, E. D., 1991, Occurrence of small colloids in sea water, *Nature 253*: 342-344.

Wells, M. L., and Goldberg, E. D., 1993, Colloid aggregation in seawater, *Mar. Chem. 41*: 353-358.

Werquin, M., Ackermann, H. -W., and Levesque, R., 1988, A study of 33 bacteriophages of *Rhizobium meliloti*, *Appl. Environ. Microbiol. 54*: 188-196.

Wilson, W. H., Joint, I. R., Carr, N. G., and Mann, N. H., 1993, Isolation and molecular characterization of five marine cyanophages propagated on *Synechococcus* sp. strain WH7803, *Appl. Environ. Microbiol. 59*: 3736-3743.

15

DIVERSITY OF NATURALLY OCCURRING PROKARYOTES

E. F. DeLong

Department of Ecology, Evolution and Marine Biology
University of California, Santa Barbara
Santa Barbara, California 93106

INTRODUCTION

Understanding the patterns of naturally occurring microbial diversity, how these patterns vary in space and time, and how they relate to ecosystem structure and function, remains a significant challenge for microbiologists. A variety of levels of microbial diversity are important and relevant from the perspective of microbial ecology, including trophic, physiological or functional diversity, intraspecific genetic diversity, or phylogenetic diversity of species and higher taxa. For the purpose of this chapter, "diversity" will be used to refer to phylogenetic variety at the level of species and higher taxa. One of the very first tasks necessary for assessing naturally occurring prokaryote diversity is to identify accurately those prokaryotic species predominant in a given habitat. This fundamental goal, representing the basis of understanding of the ecology or natural history of microbes, has rarely, if ever, been achieved for any naturally occurring microbial assemblage. We have a long way to go to catch up with the rest of biology!

SOME PROBLEMS, AND SOME SOLUTIONS

A fundamental problem preventing accurate assessment of naturally occurring microbial diversity has been the lack of appropriate methods for describing and characterizing unperturbed microbial assemblages. The problem of accurately describing naturally occurring microbial assemblages has been discussed and is reviewed elsewhere (Staley and Konopka, 1985; Pace et al., 1986), and is largely intertwined with the history and development of microbiology as a science. The heart of the problem is that pure culture techniques and cultivation methods, despite their tremendous utility and importance in the development of microbiological science, are grossly inadequate for describing complex, naturally occurring microbial assemblages. The types of microorganisms commonly recovered in standard microbiological enrichments and media often do not represent the most abundant or active components of the assemblage

Microbial Diversity in Time and Space, edited by Colwell et al.
Plenum Press, New York, 1996

125

from which they derive. For some microbial species, appropriate media and conditions for growth and recovery are simply not well developed or available (and, regrettably, may never be). Even if all possible media/growth conditions were available, the number and types of media that would be necessary to retrieve all important members of any given assemblage could be as numerous and diverse as the species themselves. Imagine imposing the requirement that botanists studying tropical rain forest biodiversity must cultivate every species identified in every field survey!

One Solution: Molecular Phylogenetic Approaches in Microbial Ecology

The insights of Norman Pace and his colleagues in the early 1980s were instrumental in removing some of the roadblocks preventing accurate descriptions of the evolutionary relationships, diversity, and ecology of natural prokaryotic assemblages (Stahl et al., 1984; Pace et al., 1986). These workers first recognized the power of molecular phylogenetics for inferring evolutionary relationships among microorganisms, even uncultivated members of complex natural assemblages. For most prokaryotic species few morphological features are available for taxonomic or phylogenetic differentiation, thus subcellular attributes (physiological, biochemical or molecular) are the characters most commonly available for descriptive purposes. Early taxonomic work relied nearly exclusively on cultivation of microorganisms to elucidate these characters. Yet in many natural assemblages, only 0.1% or less of the indigenous microbes can be readily recovered by cultivation. Pace and coworkers realized that the problem of identifying so-called "uncultivable" microbes could largely be overcome by focusing on informational macromolecules (semantides; Zuckerandl and Pauling, 1965) upon which evolutionary history is written. The approach of Pace and colleagues involved the recovery of phylogenetically informative gene sequences from nucleic acids extracted directly from naturally occurring microbial biomass. Phylogenetically informative gene sequences from the"mixed population" nucleic acids are clonally isolated, sorted, and sequenced using standard molecular biological techniques. Small subunit ribosomal RNA (ssu rRNA) genes were some of the first genes to be used as phylogenetic markers, and in many respects are still the most empirically useful. Analysis of the recovered gene sequences has allowed phylogenetic identification of naturally occurring microbes, without cultivation. Recent developments, especially the application of thermostable DNA polymerases for gene amplification, have technically streamlined the approach and made it more accessible to microbial ecologists (Sakai et al., 1988; Medlin et al., 1988; Giovannoni et al., 1990) .

Nucleic acid sequence information derived from studies of natural populations provides more than mere phylogenetic identification. It also can provide a phylogenetically-based marker, or "signature sequence" for individual species, which can be specifically tagged using nucleic acid hybridization probes. Small subunit rRNAs have proven most useful for both inference of phylogeny and "molecular tagging" and identification. One reason for their utility is the relatively large amount of evolutionary information they contain. Another unique attribute of rRNA is its relatively high intracellular abundance. This high intracellular rRNA content can provide abundant targets for phylogenetic identification of individual cells using fluorochrome-labeled rRNA probes, or "phylogenetic stains" (DeLong et al., 1989). Pace's approach of molecular phylogenetic identification, due to its avoidance of cultivation biases, has led to the recognition of novel microbial taxa ranging from new species to new phyla. Almost every study using this approach has revealed novel phylogenetic diversity.

ARCHAEBACTERIA

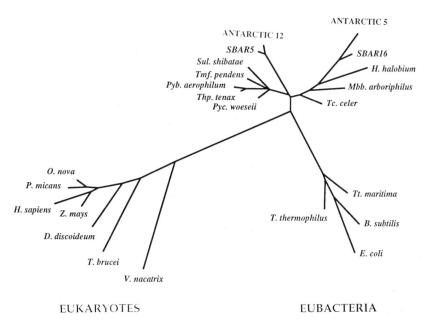

Figure 1. Phylogeny of newly discovered planktonic Archaea.

NATURAL MICROBIAL DIVERSITY: THE TIP OF THE ICEBERG?

Recent applications of molecular phylogenetics to microbial ecology provide some dramatic examples as to the extent of naturally occurring prokaryotic diversity. Bacterial phylotypes identified from a wide variety of habitats differ substantially from characterized bacteria cultured from those same environments (Giovannoni, et al. 1990; Schmidt et al., 1991; Fuhrman et al., 1992; Ward et al., 1990; Liesack and Stackebrandt, 1992; Barns et al., 1994). Only a small minority of these naturally occurring bacterial rDNA clones appear closely related to cultivated and well-characterized bacterial species from the same environments. The results indicate the dramatic and selective effect of culture-based methods historically used for recovery and cultivation of prokaryotes. They also have important implications for process-oriented microbial ecologists. Specifically, are the laboratory strains and physiological types we use to model specific biogeochemical processes really the most appropriate ones for truly understanding the pathways and processes involved in natural elemental cycles ? The answer to this question will likely vary widely, depending on the specific microorganisms, processes, and environments at issue.

Archaeal Diversity: Ignorance of Diversity in a Major Domain

An example of one surprise to come from recent phylogeny-based biodiversity studies can be found in the discovery of novel members of the Domain Archaea (sensu Woese et al., 1990). Archaea are evolutionarily unique prokaryotes, as genetically distant from "common" eubacteria as they are from eukaryotes (Woese, 1987; Figure 1). In fact, some molecular evolutionists now postulate the Archaea are more closely related to Eukarya

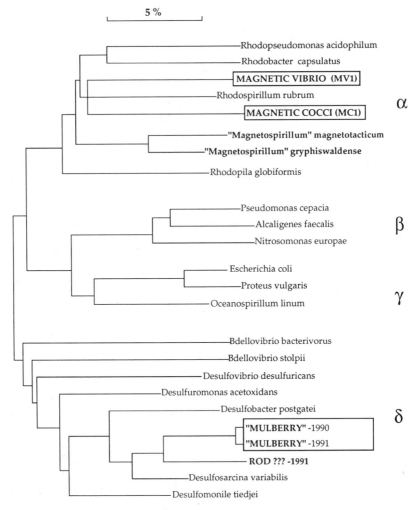

5 %

Rhodopseudomonas acidophilum
Rhodobacter capsulatus
MAGNETIC VIBRIO (MV1)
Rhodospirillum rubrum
MAGNETIC COCCI (MC1)   α
"Magnetospirillum" magnetotacticum
"Magnetospirillum" gryphiswaldense
Rhodopila globiformis

Pseudomonas cepacia
Alcaligenes faecalis   β
Nitrosomonas europae

Escherichia coli
Proteus vulgaris
Oceanospirillum linum   γ

Bdellovibrio bacterivorus
Bdellovibrio stolpii
Desulfovibrio desulfuricans
Desulfuromonas acetoxidans
Desulfobacter postgatei   δ
"MULBERRY" -1990
"MULBERRY" -1991
ROD ??? -1991
Desulfosarcina variabilis
Desulfomonile tiedjei

Figure 2. Phylogenetic relationships among magnetotactic bacteria.

(eukaryotes) than they are to Bacteria (eubacteria). Cultivated and well-characterized archaeal groups include extreme thermophiles (grow optimally at or above 80 °C), extreme halophiles (growth in very high NaCl concentrations), and methanogens (strictly anaerobic methane producers which grow at moderate to high temperatures). Because they were previously thought to thrive only in extremely hot, very salty, or strictly anoxic niches, the archaebacteria had not been considered very important in the global ecology of aerobic environments. Studies combining techiques of molecular biology and microbial ecology (Pace et al., 1986), however, have led to an unexpected observation: in cold aerobic marine environments, archaebacteria appear to be ubiquitous, abundant, and in all likelihood geochemically active components of planktonic marine microbial communities (Fig. 1; DeLong, 1992; 1994; Fuhrman et al., 1992, 1993; DeLong et al., 1994). Two new archaeal lineages were revealed in these studies (DeLong, 1992), one related to the hyperthermophilic Crenarchaeota (Fig. 1, SBAR5, ANTARCTIC12), the other to the halophile/methanogen branch, the Euryarchaeota (Fig. 1, SBAR16, ANTARCTIC5). Given that Archaea were previously thought to be confined to relatively limited and specialized habitats, these findings came as a surprise to many microbiologists.

Perhaps even more surprising was the observation that some of the highest relative abundances of planktonic archaebacteria were found in cold waters (-1.5 °C) off the Antarctic Peninsula. In these frigid marine habitats, planktonic archaebacteria comprised an estimated 34 % of the total prokaryotic biomass (DeLong et al., 1994). The most frequently encountered marine archaebacteria appear to be most closely related to a group which was previously thought to contain exclusively hyperthermophilic microorganisms, the Crenarchaeota. Yet, it now appears that members of this group have evolved to thrive at extremely low temperatures as well (Fig. 1, ANTARCTIC12). These new findings dramatically extend the growth temperature range of archaea in general, and underscore how much remains to be learned about the extent and nature of microbial diversity on the Earth - including diversity at the level of phyla within the prokaryotic Domains Archaea and Bacteria.

Diversity among Conspicuous Ecological and Physiological Types

A number of recent studies have demonstrated the apparent polyphyletic nature of several major prokaryotic physiological groupings. For example, both the sulfate-reducing bacteria and nitrifying bacteria are widely distributed within or between several major prokaryotic phyla. The sulfate reducing bacteria are found within the Gram positive Bacteria, the delta subdivision of the *Proteobacteria* (Devereux et al., 1989), and even in another domain, the Archaea. The cultured and characterized nitrite-oxidizing chemolithoautotrophs are found within the alpha, gamma, and delta subdivisions of the *Proteobacteria*, while ammonia oxidizers appear restricted to the beta and gamma subdivisions (Head et al., 1993; Teske et al., 1994). Many biogeochemically important microbial groups, defined by their physiology, appear to be broadly distributed among different phylogenetic groups. A major challenge for contemporary microbiologists is to comprehensively define and interrelate the distributional patterns of phylotype and phenotype among extant microbial species.

Another example of apparent convergent evolution can be seen in a morphologically and behaviorally conspicuous group of prokaryotes known as magnetotactic bacteria. These microorganisms contain magnetosomes, which are intracellular, iron-rich, membrane-enclosed magnetic particles. The particles confer a permanent magnetic dipole on the bacterial cell, which gives it a specific orientation in the earth's geomagnetic field as it swims. The magnetosomes of most magnetotactic bacteria contain iron oxide particles (MV1, MC1, *Magnetospirillum* sp., Fig. 2). However, some unusual colony forming magnetotactic bacteria collected from sulfidic, brackish-to-marine aquatic habitats (dubbed "Mulberry", Fig. 2), have recently been found to contain iron sulfide particles consisting of greigite (Mann et al., 1990). Phylogenetic analyses of ssu rRNA sequences showed that most iron oxide containing magnetotactic bacteria are associated with the alpha subdivision of the *Proteobacteria* (Schleifer et al., 1991; Eden et al., 1991). In contrast, iron sulfide containing magnetotactic bacteria fall within the delta *Proteobacteria* subdivision, and are specifically related to the sulfate-reducing eubacteria ("Mulberry", Fig.2; DeLong et al., 1993a). Other greigite containing rod-shaped organisms may also be specific relatives within the delta *Proteobacteria* (Fig. 2). On the whole, these findings suggest that the magnetotactic phenotype may have evolved independently several times in the course of eubacterial evolution. This hypothesis has been supported recently by the discovery of another, very large, magnetite-containing magnetotactic bacterium (*Magnetobacterium bavaricum*) which appears to fall completely outside the *Proteobacteria* lineage (Spring et al., 1993). An alternative explanation for the phylogenetic distribution of magnetosomes is that magnetotaxis was the ancestral state, but has subsequently been lost in the majority of extant *Proteobacteria*. Another possibility is that interspecific genetic transfer distributed magne-

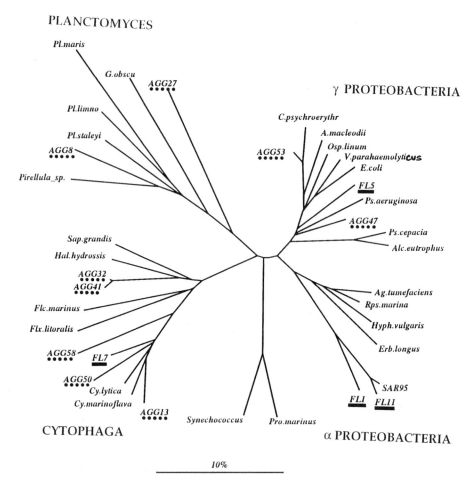

Figure 3. Diversity of particle-attached (AGG) vs. free-living (FL) bacteria.

tosome biosynthetic genes between disparate bacterial genera. For this latter possibility, a subsequent event within magnetotactic delta *Proteobacteria* (Fig 2), which resulted in replacement of magnetite by greigite, must also be postulated.

Diversity Within and Between Specific Microhabitats

The ability to assay microbial diversity without cultivation biases has great potential for providing new insight into the variety, evolution, ecology and dynamics of prokaryotic species. Can we predict population structure and composition in specific (micro) habitats of given physical and chemical features? Are there regular patterns of microbial species succession within any given changing habitat? Is there significant functional redundancy among specific assemblages? What is the relative importance of different mechanisms which result in significant functional shifts in community activity, including adaptive responses within existing bacterial species, and succession and replacement by entirely new community members? All these questions are readily addressable once accurate assessment of population structure, diversity, and its variability can be obtained. Currently, the direct linkage of phylotype with functional ecological role is tenuous at best. However, if predictable patterns of microbial diversity and succession can be followed and correlated with other environmental parameters, then the linkage between specific bacterial phylotypes and the processes which they mediate may become more robust. Clearly the future of microbial ecology will continue to be interdisciplinary in nature.

In one study of bacterial diversity in a specific habitat, the phylogenetic identity of particle-attached versus free-living marine bacteria, co-occurring in the same water mass, was compared (DeLong et al., 1993b). Samples of aggregate-associated or free-living bacteria were collected from the same water mass, their nucleic acids extracted, and small subunit ribosomal RNA genes amplified. The diversity and phylogenetic identity of associated bacterial assemblages were inferred by sequence analyses of the amplified, cloned rRNA genes. This initial study showed that ribosomal RNA genes derived from particle-associated bacteria (AGG sequences, Figure 3) were fundamentally different from those derived from free-living bacterioplankton (FL sequences, Fig. 3). The majority of rRNA types recovered from the free-living bacterioplankton (Fig. 3, FL11) were closely related to a phenotypically undescribed alpha *Proteobacteria* group, previously shown to be prevalent in waters of North Pacific and Atlantic central ocean gyres (Giovannoni et al., 1990; Schmidt et al., 1991). In contrast, the majority of particle-associated rRNA clones were closely related to *Cytophaga*, *Planctomyces*, or gamma *Proteobacteria*, within the domain Bacteria (DeLong et al., 1993b). These data indicated that specific bacterial populations, different from those which predominate in free-living bacterioplankton, develop on marine phytodetrital particles. A variety of distinctly different microbial communities may exist in "microhabitats" in areas otherwise considered to be homogeneous.

Functional significance of Phylogeny and Ecological Distributions. The high abundance of *Cytophaga* and close relatives on particulate organic material has recently been shown to occur regularly on particles collected from a variety of depths by submersible (DeLong and Fowler, in preparation) and disparate geographic locales (DeLong and Rath in preparation). Many of the *Cytophaga* and their close relatives produce one or more exoenzymes, and these microorganisms as a group degrade a wide range of polymeric compounds including proteins, polysaccharides, chitin and nucleic acids. Close relatives of some of the most abundant groups found on aggregates (*Cytophaga lytica, Saprospira grandis*) are known for their tendency to associate with surfaces, surface-associated gliding motility, and the production of hydrolytic exoenzymes. The predominant phenotypic characteristics of many of the particle-attached bacterial types, including those belonging to *Planctomycetales* and *Cytophaga*, may indicate their specific functional roles on particulate organic material. This postulate is supported to some extent by independent data on the biochemistry of marine particulate material. For example, previous studies have shown that high exoenzyme activities are often associated with particulate material in the marine environment (Smith et al., 1992). In all likelihood, this high exoenzyme activity is due to *Cytophaga* species and relatives, which appear to be predisposed to colonize particulate organic material, and have been found on all such particles so far examined (DeLong, Rath, and Fowler in preparation). The postulated linkages between resident particle-attached bacteria and particle-associated processes (particulate organic matter cycling) represent hypotheses which are now testable, due to the improved methods available for assessing naturally occurring prokaryote diversity in a relatively unbiased fashion.

CONCLUSION

We are still just beginning to gain an appreciation of the extent and nature of phylogenetic diversity within the prokaryotes. As knowledge of microbial evolution, diversity, and ecology accumulates, microbiologists will be in a position to begin to sketch the outlines of a natural history of microorganisms (Stahl, 1993). The initial phase, which is ongoing, consists largely of a discovery period delineating the nature and extent of natural microbial diversity. This initial period will, hopefully, be followed by more detailed studies, providing further insight into the pathways of prokaryote evolution, and the processes and patterns which help create and maintain that diversity which largely controls the biogeochemical balance of our planet.

ACKNOWLEDGMENTS

Thanks are extended to generous colleagues and collaborators, including Norman Pace, Richard Frankel, Dennis Bazylinski, Alice Alldredge, Raffael Jovine, and Barbara Prezelin.

I am indebted to P. Fowler, K. Y. Wu, and D. Franks for expert technical assistance. This work was supported by NOAA NURP and the West Coast National Undersea Research Program , University of Alaska Fairbanks, project No. CA93-01 and a U.S. National Science Foundation grant No. OCE-9218523.

REFERENCES

Amann R.I., L. Krumholz and D. A. Stahl. 1990. Fluorescent oligonucleotide probing of whole cells for determinative, phylogenetic, and environmental studies in microbiology. J. Bacteriol. 172: 762-770.

Barns, S. M.., R. E. Fundyga, M. W. Jeffries and N. R. Pace. 1994. Remarkable archaeal diversity detected in a Yellowstone National Park hot spring environment. Proc. Natl. Acad. Sci. USA 91:1609-1613.

DeLong E.F., G. Wickham, and N. R. Pace. 1989. Phylogenetic stains: Ribosomal RNA-based probes for identification of single microbial cells. Science 243:1360-1363.

DeLong, E. F. 1992. Archaea in coastal marine environments. Proc. Natl. Acad. Sci. USA *89* :5685-5689

DeLong, E. F., R. B. Frankel and D. A. Bazylinski. 1993a. Multiple evolutionary origins of magnetotaxis in bacteria. Science 259:803-806.

DeLong, E. F., D. G. Franks and A. L. Alldredge. 1993b. Phylogenetic diversity of aggregate-attached versus free-living marine bacterial assemblages. Limnol. Oceanog. 38:924-934.

DeLong, E. F., K. Y. Wu , B. B. Prézelin, and R. V. M. Jovine. 1994. High abundance of Archaea in Antarctic marine picoplankton, Nature 371: 695-697.

Devereux, R., M. Delaney, F. Widdel and D. S. Stahl. 1989. Natural relationships among sulfate-reducing bacteria. J. Bacteriol. 171:6689-6695.

Eden, P. A., T. M. Schmidt, R. P. Blakemore, and N. R. Pace. 1991. Phylogenetic analysis of *Aquaspirillum magnetotacticum* using polymerase chain reaction-amplified 16S rRNA-specific DNA. Int. J. Syst. Bacteriol. 41:324-325.

Fuhrman, J. F. , McCallum, K. and Davis, A. A. 1992. Novel major archaebacterial group from marine plankton. Nature *356*:148-149.

Fuhrman, J. A., McCallum, K. and A. A. Davis. 1993. Phylogenetic diversity of subsurface marine microbial communities from the Atlantic and Pacific Oceans. Appl. Environ. Microbiol. **59**: 1294-1302.

Giovannoni, S. J., T. B. Britschgi, C. L. Moyer, and K. G. Field. 1990. Genetic diversity of Sargassso Sea bacterioplankton. Nature 345: 60-62.

Head, I. M., W. D. Hiorns, T. Martin Embley, A.. J. McCarthy and J. R. Saunders. 1993. The phylogeny of autotrophic ammonia-oxidizing bacteria as determined by analysis of 16S ribosomal RNA. J. Gen. Microbiol. 139:1147-1153.

Liesack, W., and E. Stackebrandt. 1992. Occurrence of novel groups of the Domain bacteria as revealed by analysis of genetic material isolated from an Australian terrestrial ecosystem. J. Bacteriol. 174:5072-5078.

Mann, S., N. H. C. Sparks, R. B. Frankel, D. A. Bazylinski, and H. W. Jannasch. 1990. Biomineralization of ferrimagnetic greigite and iron pyrite in a magnetotactic bacterium. Nature 343:258-261.

Medlin, L., H. Elwood, S. Stickel and M. L. Sogin. 1988. The characterization of enzymatically amplified eukaryotic 16S-like rRNA-coding regions. Gene *71*: 491-499.

Pace, N. R., D. A. Stahl, D. J. Lane and G. J. Olsen. 1986. The analysis of natural microbial populations by ribosomal RNA sequences. Adv. Microbial Ecol. 9:1-55.

Sakai, R. K., D. H. Gelfand, S. Stoffel, S. J. Scharf, R. Higuchi, G. T. Horn, K. B. Mullis, and H. A. Erlich. 1988. Primer-directed enzymatic amplification of DNA with a thermostable DNA polymerase. Science *239*:487-494.

Schleifer, K. H., D. Schuler, S. Spring, M. Weizenegger, R. Amann, W. Ludwig, and M. Kohler. 1991. The genus *Magnetospirillum* gen nov., description of *Magnetospirillum gryphyswaldense* sp. nov. and transfer of *Aquaspirillum magnetotacticum* to *Magnetospirillum magnetotacticum* comb nov. Syst. Appl. Microbiol. 14: 379-385.

Schmidt, T. M., E. F. DeLong and N. R. Pace. 1991. Analysis of a marine picoplankton community by 16S rRNA gene cloning and sequencing. J. Bacteriol. 173: 4371-4378.

Smith, D. C., Simon, M., Alldredge, A. L. and Azam, F. 1992. Intense hydrolytic enzyme activity on marine aggregates and implications for rapid particle dissolution. Nature 359: 139-142.

Spring, S., R. Amann, W. Ludwig, K. H. Schleifer, H. Van Gemerden, and N. Petersen. 1993. Dominating role of an unusual magnetotactic bacterium in the microaerobic zone of a freshwater sediment. Appl. Environ. Microbiol. 59:2397-2403.

Stahl, D. A., D. J. Lane, G. J. Olsen, and N. R. Pace. 1984. Analysis of hydrothermal vent associated symbionts by ribosomal RNA sequences. Science 224:409-411.

Stahl, D. A. 1993. The natural history of microorganisms. ASM News 59: 609-613.

Staley, J. T. and A. Konopka. 1985. Measurement of in situ activities of nonphotosynthetic microorganisms in aquatic and terrestrial habitats. Ann Rev. Microbiol. 39:321-346

Teske, A., E. Alm, J. M. Regan, S. Toze, B. E. Rittmann, and D. A. Stahl. 1994. Evolutionary relationships among ammonia-and nitrite-oxidizing bacteria. J. Bacteriol. 176:6623-6630.

Ward, D. M., R. Weller, and M. M. Bateson. 1990. 16S rRNA sequences reveal numerous uncultured microorganisms in a natural community, Nature 344: 63-65.

Woese, C. R. 1987. Bacterial evolution. Microbiol. Rev. *51*: 221-271.

Woese, C. R. , O. Kandler, and M. L. Wheelis. 1990. Towards a system of organisms; Proposal for the domains Archaea, Bacteria and Eucarya. Proc. Natl. Acad. Sci. USA 87: 4576-4579.

Zuckerkandl E., and L. Pauling. 1965. Molecules as documents of evolutionary history. J Theo Biol 8:357-366

THE INTERNET AND MICROBIAL ECOLOGY

Masao Nasu

Faculty of Pharmaceutical Sciences
Osaka University
1-6 Yamada-oka, Suita, Osaka 565, Japan

INTRODUCTION

The computer and networking have promoted the development of microbial ecology significantly. In detecting bacteria at the single cell level, an important theme in this field, computers are essential for designing of gene probes, preparing primers for PCR, carrying out homology searches, estimating ORF, etc. A brief introduction to the use of Internet in microbial ecology is provided.

HISTORY

For the last three decades, computer systems have been undergoing significant change. The 1970's were a time of the "main frames" or huge host computers used in time sharing systems (TSS). The huge host computer was a symbol of the computer center. The 1980's were characterized by the personal computer. IBM-PCs and Macintosh personal computers replaced host computers for statistic analysis, data filing, database retrieval and various other purposes. IBM-PCs, Macintosh, and Windows machines were the most familiar end-user machines. However, these powerful personal computers had limitations. While they were capable of data filing on a personal basis, they were not useful for construction of databases to be shared by researchers. As for computer networking, Macintosh and IBM-PCs were suitable as terminals, but they were not applicable as server machines. The 1990's represent the era of the workstation. Before 1990, UNIX-based workstations were so predominant that they were called "engineering workstations". Now, UNIX-based workstations have rapidly become popular in many fields of science. Moreover, the development of GUI (Graphic User Interface) allows easy handling of workstations. Researchers in microbiology are no exception to these developments.

BIOLOGY AND THE INTERNET

Workstations at universities and research institutes have been organically connected to each other by the Internet, a high speed datalink. The Internet is a network of networks

Microbial Diversity in Time and Space, edited by Colwell et al.
Plenum Press, New York, 1996

135

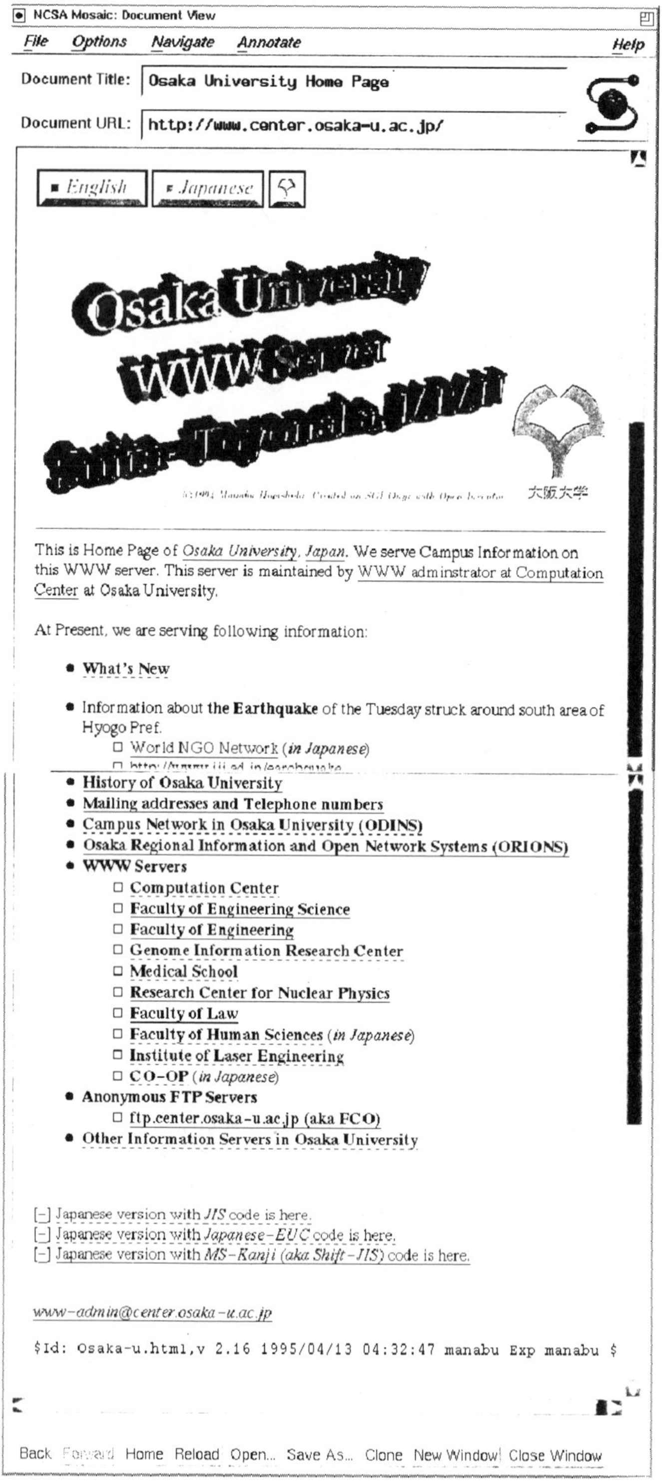

Figure 1. From the home page of Osaka University, by clicking a WWW server, the Genome Information Research Center, homology search, or estimation of ORF can be executed without much effort. (See next page).

Figure 1. (cont.)

from which volumes of resources and information are available to almost anyone. As the practicality of networking came to be widely recognized, more researchers in many fields came to utilize it routinely.

In molecular microbiology, researchers were confronted with difficulties in the management of accumulated data, and they cooperated with computer specialists to build up a system to facilitate access to extremely large biological databases.

WWW and Mosaic

World Wide Web (WWW) and Mosaic are probably the most powerful and useful services for researchers who search documents or retrieve information from distant computers connected by Internet. Moreover, these are recognized as the most familiar and commonly used at the present time.

WWW and Mosaic represent both server and client. On the Internet, a server is a huge program which provides a frame to make links or connections between a number of information resources stored in different computers and connected to the Internet. In other words, a server constructs a distributed database or data filing system. These information resources are provided by universities, research institutes, private companies, government agencies, and many other organizations. Even individuals can provide information.

The most popular server at present, WWW, deals with any type of data, text files, image files, audio files, and the like. Unlike WWW, Gopher deals only with text data. When the desired information is in text form and located in a distant computer connected by slow link, Gopher is more convenient than WWW.

In practice when accessing a server, the software, called a client, and interface, or a browser, for a server is employed. There are several client software packages, such as Mosaic, Netscape, Omniweb, etc, for different OS's.

Mosaic enables access to multiple servers, from which its name, Mosaic, originates. When a user accesses a server through a client, Mosaic, for instance, the user would not know which server is being handled. It is because Mosaic generally leads the user automatically to a home page from where, simply by clicking a word or an icon which is linked somewhere, the user is able to proceed to any file linked by the servers. From the file obtained, the user can continue the same procedure. Linked words are usually bold, highlighted, underlined, or in square brackets. If it is a graphical server/client, the links may be shown in blue or with a small square button.

The introduction of WWW brought biologists much closer to the Internet. In other words, the WWW made good use of the Internet.

Services in Biology

There are many services throughout the world for biology and biodiversity. Some of the services biologists may find beneficial include WWW Virtual Library for bioscience at Harvard University, Prot-Web and Electronic Publications for Biology at Johns Hopkins University, the World Directory of Collections of Cultures of Microorganisms (WDCM), a WFCC information resource maintained at the Institute of Physical and Chemical Research, RIKEN, Japan. World Conservation Monitoring Centre (WCMC), a joint venture between The World Conservation Union, United Nations Environment Programme, and World Wide fund for Nature provides data, including threatened species and protected area lists.

To access such services, the user starts from any server available and keeps on clicking icons or words that appear related to the aim (Fig. 1). Different services appear one after another until the intended service is obtained. By typing in the URL column of Mosaic, for example, "http://www.inbio.ac.cr", or "http://www.ftpt.br/bin21/bin21.html", information on biodiversity can be easily obtained.

How to Access

It is not easy to describe how to use the Internet, especially for those who are not familiar with computers or with computer communications because there are different computers, software, cables, etc. available to connect to the Internet. Therefore, what is stated here is in general terms. When connecting to the Internet, in practice, or if one finds difficulties along the way, it is advisable to consult an expert nearby who is an Internet user.

For those active in research, such as universities and laboratories where LAN is provided, it is usually possible to use the Internet directly with UNIX workstations, PC's or Mac's, etc.. UNIX workstations usually include Internet software. This can be confirmed by typing "telnet localhost", or "telnet rs. intermic. net". For a PC user running Windows or DOS, there are some software packages available to connect to the Internet. A Mac user needs the MacTCP newest version instead of AppleTalk.

For an individual who is not in such an environment, it is still possible to connect to the Internet by a telephone line, and by signing up for one of the providers and having an account number issued. Providers provide dialing up services, that is phone numbers, to access the Internet at cost. There are a number of providers and their rates differ widely.

CONCLUSION

Internetworking with workstations, PC's, Mac's, and other machines allows data storage, data sharing, easy retrieval, and exchange of enormous amounts of data. The Internet now allows the entire world to function as if it were within a single system. In the next few years, computer networks will become more familiar to biologists and fluorescent labeled gene probes and/or primers for PCR will become indispensable tools for single cell detection and determination of community structure. Examination of functions of microbes in natural ecosystems will be done in great detail. Analysis of genetic information will be supported by the Internet and the understanding of biodiversity will be further understood. Via the Internet, microbial ecologists throughout the world will be able to exchange data, even microscopic images, and talk easily together at any time by e-mail, leading to a genuine international communication among researchers. Virtual universities and virtual research institutes will develop and they will make valuable contributions to microbial ecology.

17

IMAGE ANALYSIS OF BACTERIAL CELL SIZE AND DIVERSITY

Kenji Kato

School of Allied Medical Sciences
Shinshu University
Asahi-3-1-1, Matsumoto 390, Japan

BACTERIA IN THE NATURAL ENVIRONMENT ARE SMALL, BUT NOT UNIFORM IN SIZE

Bacteria as they occur in nature are very small, compared with bacteria cultured under enriched conditions of the laboratory, i.e., *in vitro* . In general, planktonic bacterial cells in the natural environment are less than 1 μm in length and, therefore, biovolumes of planktonic bacteria are usually less than 0.2 - 0.3 μm³. Planktonic bacteria are very small organisms, even among planktonic constituents of aquatic ecosystems. They constitute less than 1/100 of the biovolume of small phytoplankton and less than 1/10 of that of microflagellates (Fig. 1). Compared with large phytoplankton, cell size differences reach more than thousands-fold. However, as planktonic bacteria are plentiful in water, in the order of 10^5 to 10^7 per ml (Van Es and Meyer–Reil, 1982), their overall biovolume in aquatic ecosystems can be significant. In marine systems, small bacteria, with a diameter of ca. 0.2 - 0.6 μm usually predominate, but, even among these bacteria, significant variation in size exists. Therefore, the bacterial biomass of a given population cannot be estimated accurately from a roughly estimated mean value, even if the diameters range from 0.2 to 0.6 μm, since such differences can lead to a ten-fold increase in biovolume. When bacteria are observed under an epifluorescence microscope, very large cells are often found, even in samples of pelagic marine waters. Thus, it is necessary to determine the cell size of bacteria precisely. The development of a personal computer-assisted image analyzer, connected to an epifluorescent microscope via a charge-coupled device (CCD) camera, has enabled the size of small bacterial cells to be measured automatically (Sieracki et al., 1985; Bjørnsen, 1986).

IMAGE ANALYSIS VERSUS FLOW CYTOMETRY

At present, there are two methods for measuring the size of hundreds of bacterial particles automatically using epifluorescence: image analysis (IA) and flow cytometry (FCM). For FCM, no sample preparation is necessary and numerous particles in suspension

Microbial Diversity in Time and Space, edited by Colwell et al.
Plenum Press, New York, 1996

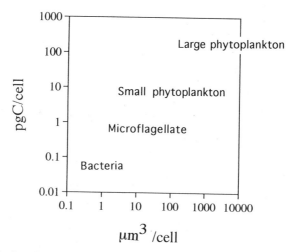

Figure 1. Size distribution of lower trophic level plankton. Figure is drawn from the data of Verity et al. (1992) and also unpublished data.

can be measured very quickly. In contrast, a "Präparat" must be prepared for IA. This process traps the three-dimensionally suspended bacteria in a two-dimensional filter. However, it is possible to estimate the size of each cell in an aggregate from the IA operating digitized image, whereas in FCM aggregated cells are treated as a single particle. To count bacteria, there is no significant difference between these techniques. However, for particle size information, IA provides simultaneous geometric information. Furthermore, absolute size can be obtained from IA measurements. Thus, since it is not necessary to separate bacterial particles, using a specially equipped cell sorter, IA is more useful than FCM for ecological measurements of bacterial cell size. In addition, an IA facility is not as expensive, e.g. ca. US$ 7,000 including the microscope, as the FCM apparatus.

Although satisfactory software and hardware are available for IA, several fundamental procedures must be operated manually very carefully. The most important step is to obtain as bright and clearly contrasted an image as possible. To achieve this, a very sensitive 3CCD camera (e.g. Olympus-Ikegami ITC-380M) should be employed, choosing a light pass from red, green, or blue, instead of a synthesized one, according to the color of the sample. The second most important step is to fix the gray level, which is done using an analog image

Table 1. Comparison of flow cytometry and image analysis

Function	Flow cytometry	Image analysis
Counting	Yes	Yes
Sizing	Yes, but does not provide absolute values	Yes
Information provide concerning cell shape	No	Yes
Distinguishes individual cells from aggregates	No	Yes
Sorting	Yes	No

monitor, as well as a digitized monitor. Natural samples contain at least several dividing cells, which, by virtue of their distinct shape, can be used as an active marker to assist in finding a suitable gray level (see details in the technical discussion by Sieracki et al. (1985, 1989), Bjørnsen (1986), and Schröder et al. (1991). The system employed in my laboratory consists of an Olympus BH-2 epifluorescent microscope equipped with an UVPL 100 objective lens and an image analyzer (Olympus-Avio XL-500). This system enables 45 parameters of length, width, volume, roundness, and peripheral length of each particle to be measured simultaneously. The biovolume is estimated by vertical integration of the value of a thin cylinder at each pixel: $CYV=(S(w_i/2)^2x\ \pi)\delta d$, where w_i is the length at each vertical pixel of the particle and δd is the absolute height of one pixel (0.08 μm in the system described here).

DAPI VERSUS AO

Two epifluorescence dyes are widely used to stain bacteria, 4'6-diamidino-2-phenylindole (DAPI) and acridine orange (AO). DAPI is a highly specific stain for DNA under a wide range of conditions (Porter and Feig, 1980). When excited with light at a wavelength of 365 nm, the DNA-DAPI complex fluoresces bright blue. In contrast, the staining specificity of AO varies with staining and observation conditions. Differential staining of double- and single- stranded nucleic acids can be achieved by AO (Darzynkiewicz, 1990), which intercalates into double-stranded nucleic acids and, when excited with blue light, fluoresces green, with maximum emission at 530 nm, whereas, interaction of AO with single-stranded nucleic acid results in condensation and subsequent agglomeration of the product. The wavelength of the luminescence of the condensed product is at the red end of the spectrum and maximum emission occurs at 640 nm. Thus, for precise estimation of RNA content, samples must be treated with chelating agents to denature selectively any double-stranded RNA. In the standard method, developed by Hobbie et al. (1977), a water sample is simply stained with 0.01% AO (final concentration) and the bacteria are trapped on a 0.2-μm pore-sized Nuclepore filter. The sample is excited at 490 nm, which reduces autofluorescence significantly. The DAPI staining procedure is very similar and the technique of Porter and Feig (1980) is followed.

Fig. 2a shows cell size distributions of *Escherichia coli* stained with AO or DAPI. The difference between estimated volumes is very pronounced. The volumes estimated using DAPI are highly concentrated between 0.2 to 0.4 μm³, whereas those estimated using AO show a much wider range, over 1 μm³. The mean measured cell length was 1.79 ± 0.16 μm by DAPI, whereas it was 2.54 ± 0.20 μm by AO. However, the width did not differ significantly, 0.29 ± 0.01 μm and 0.30 ± 0.02 μm for DAPI and AO, respectively. These results suggest that the AO-stained area was larger than that of DAPI, which stained DNA specifically. Although sample preparation for these measurements is simple, as described above, clear differences between results obtained with these staining procedures were observed.

The same comparison was carried out using a natural water sample. Fig. 2b shows results of IA of a water sample collected from Lake Kizaki, a mesotrophic lake. Before carrying out the staining procedure, the water was filtered through a 1-μm pore-sized Nuclepore filter to remove detritus and eukaryotic cells. Similar results for the lake bacterial assemblage were obtained. The estimated particle volume, after DAPI staining, was predominantly 0.05 - 0.5 μm³, whereas that obtained by AO staining showed a much wider range, up to 1.5 μm³ and greater. These results suggest that measurement of cellular parameters obtained using DAPI staining does not include a significant portion of the RNA, although it is not confirmed yet that AO stains the whole cell.

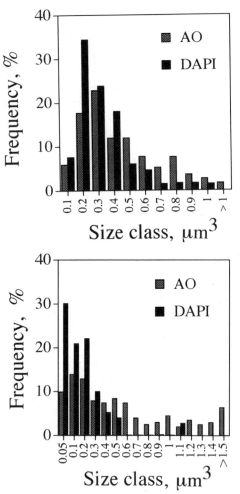

Figure 2. Size spectrum measured for DAPI and AO stained area of a) *Escherichia coli,* b) Lake Kizaki.

CHANGES IN CELLULAR SIZE WITH GROWTH PHASE

Using the technique of Hiraga et al. (1989), which involves smearing the bacterial cell suspension on a glass slide, followed by staining with DAPI, a very clear image of the DAPI-stained area is obtained, which enables precise measurements to be made. Furthermore, the shape of the nucleoid is clearly visible. When the cell size is measured using this technique, differences in size between DAPI- and AO-stained cell parameters are even more obvious.

The size distribution of cell parameters of *Escherichia coli* K-12 wild type, cultured in nutrient broth (yeast extract 0.1%, polypeptone 0.1%, NaCl 0.05% w/v, pH 7.2) at 37°C with shaking at 100 rpm, was obtained by staining with AO and DAPI during different growth phases (Table 2). Although each population examined was mixed, the size of the AO-stained area decreased markedly as the growth phase progressed from log to stationary and under starved conditions for 2 days in natural clear river water. These results suggest that the amount of single-stranded RNA, at the least, decreased sharply from log to stationary phase.

Table 2. Difference in cell parameters for *Escherichia coli* K-12, wild type in different phases of growth

Growth phase	AO stain size	DAPI stain size
Logarithmic	0.86 ± 0.64[a] (142)[b]	0.27 ± 0.14[a](112)
Late logarithmic	0.75 ± 0.63 (112)	0.12 ± 0.12 (101)
Stationary	0.64 ± 0.57 (117)	0.073 ± 0.10 (173)
Starved[c]	0.53 ± 0.35 (133)	0.11 ± 0.10 (70)

a: μm^3
b: Measured cell number
c: Bacteria were transferred to clear river water and incubated for 2 days, after which the measurement was made.

The difference between stationary phase and starved cell RNA content would be expected to be large, when starvation is prolonged. The difference between the largest value obtained during log phase and the smallest under conditions of starvation was about 40%.

The size of the DAPI-stained area, i.e., the DNA, changed only approximately two-fold, which may represent the cell cycle distribution of the majority of cells in each phase. The size of the DAPI-stained area amounted to only 10 to 30% of that stained by AO throughout the growth phase.

When bacterial cell sizes are discussed in the context of their ecological function involved in the flux of matter, knowledge of the amount of cellular carbon and/or nitrogen possessed by each cell is fundamental. Cellular carbon and nitrogen content relate primary to protein synthesis. Thus, the single-stranded RNA content may be an ecologically important functional parameter and AO staining is recommended for this purpose.

AN EXAMPLE OF DIVERSITY IN BACTERIAL CELL SIZE IN A DROP OF WATER AND A CASE OF VERTICAL CHANGE

Fig. 3 shows bacterial cell size distribution in a drop of water collected from a eutrophic lake, Lake Suwa. The predominant cell volume ranged from 0.2 to 0.4 μm^3 and

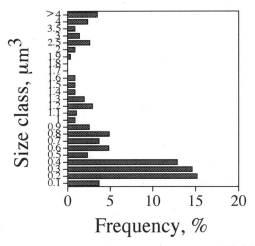

Figure 3. Size distribution of cells in a drop of water collected from a eutrophic lake, Lake Suwa, September 30, 1990.

about 50% of the total number of bacterial cells were in this size range. However, giant cells, with volumes of over 2.5 μm³ accounted for more than 10%, by number, and the total biovolume was almost twice the integrated value for the smaller sized population. Therefore, the ecological activity of these large cells, although they may be in the minority, is crucial to considerations of matter flux.

Table 3 shows the vertical change in bacterial cell size in Lake Kizaki. The mean cell size of the bacterial population decreased from the surface to 5 m below the surface, but increased again toward the bottom. At the surface, the number of bacteria was high, 2 - 3 x 10⁶ cells/ml, whereas there were less than 10⁶ cells/ml below the thermocline (Fig. 4). Therefore, the numerical bacterial biomass at the surface and on the bottom differed several-fold, but, when the mean cell size was taken into consideration, the difference between these two systems was only about two-fold. Although large bacteria inhabiting regions near the bottom probably affected the mean cell size, the data suggest that bacterial cell size cannot be assumed to be the same, even in the water column.

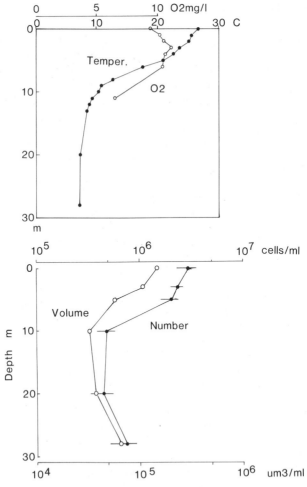

Figure 4. Vertical profile of temperature and dissolved oxygen (a), and bacterial number and biomass, calculated using mean cell volume for each depth (b).

Table 3. Vertical size distribution of bacteria in Lake Kizaki[a]

Depth	Minimum	Maximum	Mean	Standard deviation	Number of cells measured
0[b]	0.0038[c]	0.435	0.050	0.061	100
3	0.0049	0.232	0.046	0.051	83
5	0.0049	0.110	0.029	0.023	93
10	0.0049	0.546	0.067	0.098	77
20	0.0049	0.726	0.085	0.113	82
28	0.0043	0.605	0.085	0.113	127

a: Water sample collected August 22, 1991. b: metersc: $\mu m3$

When large-scale matter flux, with respect to the global environment, is considered, it must be borne in mind that the most rough estimate affects the final result. The biomass of the most abundant constituents of every ecosystem, the bacteria, should be estimated precisely, and IA is an useful tool for measuring individual bacterial cell sizes.

ACKNOWLEDGMENTS

This study was supported, in part, by The Nippon Life Insurance Foundation of 1989 and a Grant-in-Aid for Scientific Research (No. 05680445) from the Ministry of Education, Science and Culture of Japan.

REFERENCES

Bjørnsen, P.K., 1986, Automatic determination of bacterioplankton biomass by image analysis, Appl. Environ. Microbiol. 51:1199-1204.

Darzynkiewicz, Z., 1990, Differential staining of DNA and RNA in intact cells and isolated cell nuclei with acridine orange, In: Methods in Cell Biology, Vol. 33:285-298, Academic Press.

Hiraga, S., Niki, H., Ogura, T., Ichinose, C., Mori, H., Ezaki, B., and Jaffe, A., 1989, Chromosome partitioning in *Escherichia coli* : novel mutants producing anucleate cells, J.Bacteriol. 171:1496-1505.

Hobbie, J.E., Daley, R.J., and Jasper, S., 1977, Use of nuclepore filtesr for counting bacteria by fluorescence microscopy. Appl.Environ. Microbiol., 33:1225-1228.

Porter, K.G. and Feig, Y.S., 1980, The use of DAPI for identifying and counting aquatic microflora. Limnol. Oceanogr. 25:943-948.

Schröder, D., Krambeck, C., and Krambeck, H-J., 1991. How to count and size fluorescent microbial plankton with digital image filtering and segmentation. Acta Stereologica,10/1.

Sieracki, M.E., Johnson, P.W., and Sieburth, J.M., 1985. Detection, enumeration, and sizing of planktonic bacteria by image-analyzed epifluorescence microscopy, Appl. Environ. Microbiol.,49:799-810.

Sieracki, M.E., Reichenbach, S.E., and Webb, K.L., 1989. Evaluation of automated threshold selection methods for accurately sizing microscopic fluorescent cells by image analysis. Appl. Environ. Microbiol. 55:2762-2772.

Van Es and Meyer–Reil, L., 1982. Biomass and metabolic activity of heterotrophic marine bacteria, In Advances in Microbial Ecology, 111-170, Academic Press.

Verity, P.G., Robertson, C.Y., Tronzo, C.R., Andrews, M.G., Nelson, J.R., and Sieracki, M.E.,1992. Relationship between cell volume and the carbon and nitrogen content of marine photosynthetic nanoplankton. Limnol. Oceanogr. 37:1434-1446.

18

MICROBIAL DIVERSITY AND THE CYCLING OF NITROGEN IN SOIL ECOSYSTEMS

Tohru Ueda [1] and Kazuyuki Inubushi [2]

[1] Department of Agricultural Chemistry
Kyushu University
Hakozaki 6-10-1, Fuk uoka 812, Japan
[2] Faculty of Horticulture
Chiba University
Matsudo 648, Matsudo, Chiba, 271, Japan

INTRODUCTION

Nitrogen (N) is a critical element in crop production for both paddy and upland soil ecosystems. In this chapter, we describe the role of microbial biomass and its activity in Nitrogen(N) cycling supporting rice growth in the paddy soil ecosystem. We also focus on the genetic diversity of N_2-fixing bacteria in rice roots. Finally, we discuss the genetic diversity of both soil bacteria, in general, and rhizobia, specifically.

MICROBIAL BIOMASS AND DIVERSITY IN THE CYCLING OF NITROGEN IN PADDY SOIL ECOSYSTEMS

Paddy fields, or wetland rice fields, cover 145 million ha throughout the world, mainly in Asian countries, which represent slightly more than 10% of the total global cultivated areas. However, paddy fields provide rice as food staple for more than 2 billion people, a number that is increasing rapidly. Paddy fields are one of the most sustainable crop systems, since rice can take up more than 60% of its N from soil organic N, even under modern cultivation and fertilization conditions. There are two important processes to maintain soil organic N. Firstly, N immobilization by soil microorganisms, which then provides N to the plant through N-mineralization, and secondly from the atmosphere by biological N_2 fixation.

Microbial Biomass and Its Activity in Waterlogged Soils

N-mineralization is carried out by many heterotrophic soil microorganisms in the microbial biomass. We measured microbial biomass in paddy soils and its activity, as [15]N

Microbial Diversity in Time and Space, edited by Colwell et al.
Plenum Press, New York, 1996

turnover rate and ATP metabolism. Microbial biomass N was determined first by the chloroform fumigation-incubation method in paddy fields in the Philippines (Inubushi and Watanabe, 1986). Nitrogen-15 labelled fertilizer was applied only at the beginning of the experiment, to determine ^{15}N in microbial biomass and plant uptake.

Microbial biomass decreased only slightly at the beginning and then remained fairly constant until the end of the crop season; even the rice plant took up N actively (Fig. 1). However, rice ^{15}N content initially increased then decreased. Nitrogen-15 content in both the biomass and plant became more similar, indicating that the biomass became a direct source of N late in plant growth. The turnover rate of biomass was calculated from the dilution of ^{15}N in the biomass, which was about 20 days, much faster than reported for temperate upland soil (Jenkinson and Rayner, 1977)

In laboratory experiments, biomass was also measured by ATP (Inubushi et al., 1989). The microbial biomass stayed constant during aerobic incubation, but decreased slightly during anaerobic incubation (Fig. 2). Furthermore, when permanently aerobic arable or grassland soils were incubated under waterlogged conditions, ATP declined much faster. These results indicate that biomass ATP concentration declines under anaerobic conditions. We also measured ADP and AMP to calculate AEC (adenylate energy charge) an indication of microbial activity (Powlson, 1994). When soil samples were incubated anaerobically, AEC decreased from 0.9 to about 0.5 or less, but recovered quickly when the soil was aerated, suggesting that microorganisms in waterlogged soil can maintain their ability to respond

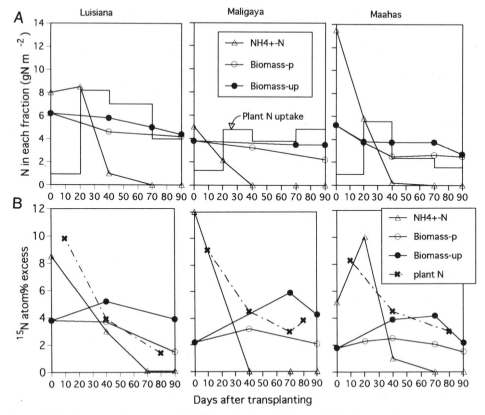

Figure 1. (a) Changes in the content of NH4[+]-N and microbial biomass N in rice-planted (p) and unplanted (up) paddy soils and N uptake by rice plant and (b) ^{15}N abundances in three kinds of Philippine soils.

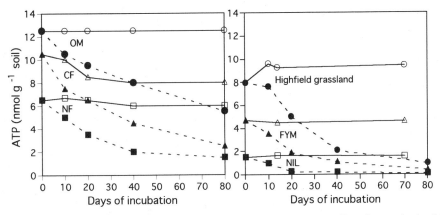

Figure 2. ATP content of three paddy soils collected in Japan (left), two arable soils and a grassland soil from England (right) incubated under aerobic (solid lines) or anaerobic (dashed lines) conditions. OM; organic manure, CF; chemical fertilizer, NF; no fertilizer, FYM; farmyard manure, NIL; no fertilizer.

quickly to oxygen. The microbial community structure in paddy soils also needs to be investigated by analyzing the molecular diversity of 16S rRNA extracted from soil.

N_2-Fixing Bacterial Diversity Detected in Rice Roots by Analysis of *nif* Gene Sequences

In flooded soils, nitrogen is available to the rice plant even in fields that have been planted for many years without fertilizer application. The long maintenance of soil fertility is considered to be due to biological nitrogen fixation. It has been suggested that N_2 fixation in the rice root zone is associated with the activity of N_2-fixing heterotrophic bacteria that inhabit the rice rhizosphere (Watanane and Barraquio, 1979).

To demonstrate the extent of phylogenetic diversity of diazotrophic bacteria associated with rice roots, we characterized phylogenetically the *nif* gene sequences obtained by PCR amplification of mixed organism DNA extracted directly from rice roots, without culturing. The *nifD* gene, encoding alpha subunit of dinitrogenase, is one of the most informative genes among the *nif* genes. About 450 bp fragments of *nifD* were amplified from the root DNA. Products were purified and the *nifD* fragments were cloned into pT7BlueT vector to construct a *nifD* library, using *E. coli* . Sixteen cloned *nifD* genes chosen at random from the library were sequenced, i.e., clones D-RIC1 to D-RIC16.

Fig.3 shows an evolutionary tree constructed from the data, analysed by the NJ method (Ueda *et al.*, 1995c). This phylogenetic tree clustered archaea with gram-positive bacteria of low G+C content, gram-positive bacteria of high G+C content, and proteobacteria, successively (slightly) more distantly related in a rooted tree. In this respect, the NifD tree is congruent with the 16S tree (Woese, 1987). Although some of the branching order of the groups differed to some extent, it is important to note that the outline of the groups established and the deeper branching order were largely congruent. Therefore, it can be assumed that a bacterium with a sequence of the *nifD* gene which falls in some known cluster might be related to the bacteria in that cluster. The analysis indicated the presence of seven novel NifD types, which implies at least seven components in the diazotrophic community of the rice root. Among the seven groups described above, the D-RIC1, D-RIC3, and D-RIC6 clusters were the largest. D-RIC1 and D-RIC3 clusters, with 9 clones, branched within the

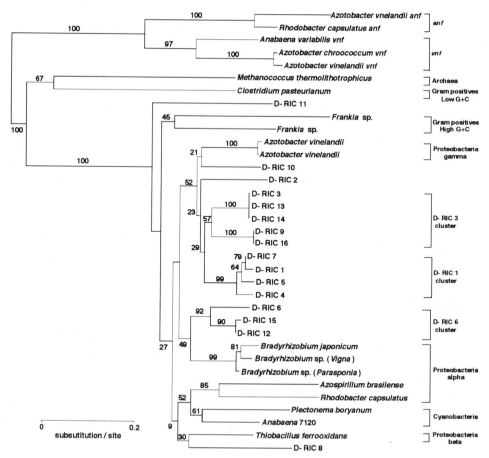

Figure 3. Phylogeny based on *NifD* protein sequences, including rice clones calculated from estimated pairwise amino acid substitutions per site by the NJ method (Ueda *et al.*, 1995c). The percentage of 1,000 bootstrap resamplings that support each topological element is indicated.

radiation of gamma-proteobacteria. These data indicate a high diversity of *nifD* gene lineages among related gamma-proteobacteria. The D-RIC6 cluster of three clones was related to the bradyrhizobia, which are alpha-proteobacteria. In order to elucidate how prevalent these organisms are in rice roots, hybridization of specific oligonucleotide probes will be needed. A study using *nifH* sequences instead of *nifD* sequences has also been done (Ueda *et al.*, 1995b).

MICROBIAL DIVERSITY AND THE CYCLING OF NITROGEN IN UPLAND SOIL ECOSYSTEMS

Genetic Diversity of Soil Bacteria in Soybean Fields

We discuss here phylogenetic characterization of small subunit rRNA gene sequences, obtained by PCR amplification of mixed population DNA extracted directly from soil in a soybean field without culturing the organisms (Ueda *et al.*, 1995d). As shown in

Fig.4, results of phylogenetic analysis of data for 17 soil clones (FIE) by the NJ method shows that the soil sample contained broadly diverse prokaryotes: a clone related to archaea; a clone related to gram-positive bacteria of high G+C content; two clones of green sulfur bacteria; and four clones of proteobacteria. Interestingly, nine soil clones were not in the clusters of any previously reported bacterial groups, which suggests they belong to members of novel groups of bacteria. In addition, the archaeal sequence, FIE16 (D26266), was found to be phylogenetically similar to ANTARCTIC12 (U11043), a clone obtained from surface waters of Antarctica (DeLong et al., 1994). Their occurrence in both the ocean and soil suggests a global distribution of this archaeal group. In conclusion, rRNA gene sequences recovered from soil biomass document the occurrence of many more bacterial lineages than have been recognized previously through cultivation.

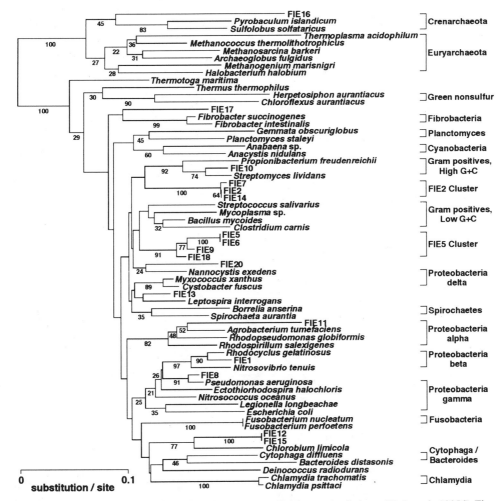

Figure 4. NJ tree, including members of all major bacterial divisions and soil clones (Ueda et al., 1995d). The percentage of 1,000 bootstrap resamplings that support each topological element is indicated.

Coevolution in Plasmid-legume Interactions

Plants coexist with a large number of soil bacteria. Rhizobia-legume interaction stands out from other plant-microbe interactions as one in which a true developmental mutualism occurs. The ability of rhizobial strains to form effective nodules, in which they reduce atmospheric nitrogen to ammonia and supply the plant with nitrogenous compounds, is limited to certain host plants. For example, *Vicia* and *Pisum* spp. are host plants for *Rhizobium leguminosarum* bv. viciae, *Phaseolus* spp. are hosts for *R.leguminosarum* bv.phaseoli, *Trifolium* spp. are hosts for *R.legumunisarum* bv.trifolii, *Medicago* spp. are hosts for *R. meliloti*, *Glycine* spp. are hosts for *Bradyrhizobium japonicum*, and the tropical legume *Sesbania rostrata* is the host for *Azorhizobium caulinodans*. These events occur when the rhizobia respond to the presence of specific plant flavonoids that stimulate the coordinate expression of bacterial nodulation genes (*nod* genes). The *nod* genes, in turn, encode enzymes involved in the synthesis of Nod factors, which act as determinants of host specificity and lead to formation of nitrogen-fixing root nodules. The *nodABC* have been characterized as common *nod* genes, essential for nodulation to occur. In *Rhizobium* species, a majority of genes coding for nodulation and nitrogen fixation are located on high molecular weight plasmids, Sym plasmids, and it is widely believed that Sym plasmids have transferred horizontally during the evolutionary process (Young and Johnston, 1989).

In order to understand host specificity of the rhizobia, we determined the partial *nodC* sequences of 10 representative strains of rhizobia and constructed evolutionary trees from the deduced amino acid sequences. We have also constructed phylogenetic trees from NodA protein sequence data, using public databases, showing that the trees have the same topology as the tree constructed from NodC protein data. Furthermore, the NodC tree has also the

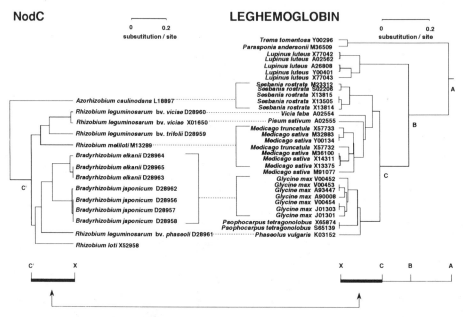

Figure 5. Phylogenetic analysis of NodC protein and leghemoglobin by UPGMA. The letters (A, B, C, C′) show branch points with corresponding divergence times. The first ancestors of legumes are considered to have appeared 100 to 120 Myr ago (Ochman and Wilson, 1987), which time might correspond to a point between A and B. Assuming that coevolution has occurred in the plant-plasmid interaction, the time scale (C-X) in leghemoglobin should roughly correspond to C′-X in NodC, within an order of magnitude.

same topology as the NodD tree, constructed by other workers (Young and Johnston, 1989). This analysis suggests that *nodA, nodC,* and *nodD* genes have evolved in a similar manner, which agrees with the fact that the *nodABCD* genes are linked. Interestingly, these coding sequences provide a phylogenetic tree similar to the leghemoglobin phylogeny of host plants, as shown in Fig.5. These observations, together with the fact that strains of *Rhizobium* species have nodulation genes on the Sym plasmids, suggest that the evolution of common nodulation genes may be linked to legume evolution and speciation. In other words, mutual adaptations of the plasmids and the legume have become refined in the course of their association and, thus, common *nod* genes and their host plants appear to radiate in parallel. This pairwise coevolution may be related to the interaction of the structure of Nod factors and corresponding receptor proteins in legumes, on an evolutionary time scale. To the best of our knowledge, our experiments provide the first molecular evidence that coevolution might have occurred in a plant-microbe interaction (Ueda *et al.*, 1994; 1995a), because in order to propose "coevolution", we should analyze not only bacterial genes but also plant genes. Assuming coevolution has occurred in plasmid-legume interactions, the data shown in Fig.5 suggest that the evolutionary rate of NodC protein during this period can be roughly calculated as 1% per site per 1.3-2.5 million years, within an order of magnitude, whereas the rate of leghemoglobin is calculated as 1% per site per 2.2-4.1 million years (Fig.5). Moreover, these data suggest not only that the partial sequence of *nodC* can be used for identification of rhizobia at the species level, but also that phylogenetic analysis of common *nod* genes may be significant in assessing phylogenetic relationships with the host plant.

Thus, results of the molecular ecological analyses reported here can provide new insight into microbial diversity in time and space.

REFERENCES

DeLong, E. F., Wu, K. Y., Prezelin, B. B., and Jovine, R. V. M., 1994. High abundance of Archaea in Antarctic marine picoplankton, *Nature* 371:695-697.

Inubushi, K. and Watanabe, I., 1986. Dynamics of available nitrogen in paddy soils II, *Soil Sci. Plant Nutrit.* 32: 561-577.

Inubushi, K., Brookes, P. C. and Jenkinson, D. S., 1989, Adenosine 5'-triphosphate and adenyrate energy charge in waterlogged soil, *Soil Biol. Biochem.* 21: 733-739.

Jenkinson, D. S. and Rayner, J. H., 1977. The turnover of soil organic matter in some of the Rothamsted classical experiments, *Soil Sci.* 123: 298-305.

Ochman, H., and Wilson, A. C., 1987. Evolution in bacteria: evidence for a universal substitution rate in cellular genomes, *J. Mol. Evol.* 26:74-86.

Powlson, D. S., 1994. The Soil Microbial Biomass: Before, beyond and back, *Beyond the Biomass*, 3-20, John Wiley & Sons Publication, Chichester.

Ueda, T, Suga, Y., Ashizuka, Y., Yahiro, N., Yamamura, M., and Matsuguchi,T., 1994. Evolution of nod genes and the Host Specificity (part1), Abstr. of the 1994 Meeting (March), Jpn. Soci. Soil Sci. Plant Nutri.. 40:49. (in Japanese)

Ueda, T, Suga, Y., Yahiro, N., and Matsuguchi, T., 1995a. Phylogeny of Sym plasmids of rhizobia by PCR-based sequencing of a nodC segment, *J. Bacteriol.* 177: 468-472.

Ueda, T, Suga, Y., Yahiro, N., and Matsuguchi, T., 1995b. Remarkable N$_2$-fixing bacterial diversity detected in rice roots by molecular evolutionary analysis of *nifH* gene sequences., *J. Bacteriol.* 177:1414-1417.

Ueda, T, Suga, Y., Yahiro, N., and Matsuguchi, T., 1995c. Genetic Diversity of N$_2$-fixing bacteria assiciated with rice roots by molecular evolutionary analysis of a *nifD* library, *Can. J. Microbiol.* 41:235-240.

Ueda, T, Suga, Y., and Matsuguchi, T., 1995d. Molecular phylogenetic analysis of a soil microbial community in a soybean field, *Eur. J. Soil Sci.* , 46:415-421

Watanabe, I. and Barraquio, W., 1979. Low levels fixed nitrogen required for isolation of free-living N$_2$-fixing organisms from rice roots, *Nature* 277:565-566.

Woese, C. R., 1987. Bacterial evolution. Ann Rev. Biochem. 51:221-271.

Young, J. P. W. and Johnston, A. W. B. 1989. The evolution of specificity in the legume-rhizobium symbiosis. *Trends Ecol. Evol.* 4:341-349.

19

THE ROLE OF MICROBIAL DIVERSITY IN THE CYCLING OF ELEMENTS. (SUMMARY OF WORKSHOP)[*]

Rita R. Colwell

University of Maryland Biotechnology Institute
4321 Hartwick Road, Suite 550
College Park, Maryland 20740

Prokaryotes are the oldest and most diverse group of organisms, encompassing a wide array of adaptations to extreme environmental conditions. The prokaryotes, including both *Archaea* and *Eubacteria*, comprise a large and versatile gene pool which can be translated to include highly versatile physiological responses to the environment, resulting in ability to grow within an extraordinarily-wide spectrum of environmental conditions. The communal metabolism of microorganisms is largely responsible for global cycling of elements, resulting in evolution of the Earth's biosphere throughout the living history of the Earth and in maintaining dynamic stability of the present biosphere. Microbial activities are also closely associated with present global climatic changes, both accelerating and slowing these processes.

Only a small number of prokaryotes are available in axenic culture in culture collections. The number is estimated to be *ca.* 4,000 and probably comprises less than 0.5 percent of the microbial species living on the Earth, at most. One of the reasons for this limited collection is the difficulty experienced in cultivating organisms from natural environments. They often exist under very low nutrient conditions, i.e., starvation, and resist cultivation in the usual microbiological media. Also, many live in close associations, e.g., coenoses, with other organisms. Another reason is the complexity of processes in nature. Diverse microorganisms often are very similar in cell morphology and in functional activities, taking part in processes which, in the natural environment, are integrated to proceed simultaneously. We do not yet understand how to provide conditions for culture of individual species. Consequently, the complexity of microbial environments has caused scientists to view microorganisms as comprising a "black box", when studying the cycling of matter, although scientists are fully aware of the cardinal importance of the activities of individual microorganisms. This complexity of microbial communities and their roles in various

---

[*] In the discussions which follow (chapters 19 and 20), it is important to note that the remarks (questions and comments) were not prepared in written form by those participants, but represent summaries of the informal workshop discussion.

Microbial Diversity in Time and Space, edited by Colwell et al.
Plenum Press, New York, 1996

processes, as well as the difficulty in defining these communities often cause some scientists to doubt whether full understanding of microbial communities will ever be achieved. The main approach to date has been to analyze isolated functions of microorganisms in these processes. Isolation and study of individual microorganisms need to be achieved if understanding, and possible remediation of, certain environments can be undertaken.

We emphasize the necessity to employ innovative approaches to obtain novel organisms in axenic culture, as well as to define co-cultures of key microorganisms. Such approaches are needed, even though use of molecular techniques will allow us, at the present time, to understand phylogenetic relationships among uncultured microbial communities and, thereby, study regulation of key metabolic enzymes in microbial communities as a function of the conditions of the environment.

The volume of molecular data in databases is growing exponentially and serves as a baseline for assessment of molecular microbial biodiversity. These databases should be compared with physiological data for axenic cultures of bacteria and ecophysiological data on microbial community metabolism to obtain better understanding of the role of microbial diversity in the cycling of elements.

An integrated effort should be undertaken to obtain an understanding, at the molecular level, of environmental signals triggering functioning of key enzymes, both in pure culture and in selected microbial communities, as well as in biochemical and physiological studies of biotransformation of elements, biogeochemistry of element cycling and, finally, in determining global effects of these processes.

It is suggested that updated microbial cycles of elements be constructed, defining gaps in present understanding of the cycling of elements. Element cycling should be dealt with considering the following:

1. Cycles of individual elements, e.g., carbon, nitrogen, sulfur, phosphorus, iron, manganese, etc. In each cycling process, the analysis of microbial communities, key microorganisms, and their functioning under *in situ* conditions should be investigated. It is both necessary and important to understand key processes which are presently not fully understood, e.g., anaerobic ammonia and methane oxidation and aerobic sulfate reduction.

2. In natural environments, different key processes often proceed by integration of various cycles of elements. An example is oxidation and reduction of iron and manganese in relation to nitrogen and sulfur cycling. A study of the dynamic control system for integrating these processes is necessary.

3. Selection of key stratified microbial communities for modeling studies aimed at the functioning of element cycling, using molecular, microbiological, biochemical, and biogeochemical tools.

4. Understanding differences in element cycling in important biotopes: the deep ocean and shallow waters, soils of arid regions, tropical environments, etc.

5. Studies of element cycling in extreme environments, e.g., extreme temperature, high salinities, extreme pH, and high pressure.

6. Combining information on microorganisms and their functions into available computer databases. Information for various microorganisms functioning in cycling of matter should include molecular data in databases, where available, via computer networks. The data should include information on axenic cultures of microorganisms and nonaxenic microbial associations, as well as molecular data for these microorganisms.

7. Studies of the cycling of matter in relation to changing environments. Human activities have caused significant transformations of natural environments. Microbial processes involved in changing forests, arid soils, and lakes and coastal

seas should be studied. Knowledge obtained from these studies form the basis for bioremediation of these environments.

Microbial diversity, in terms of species and functioning, must be fully understood in global cycling of matter. Construction of generalized schemes that incorporate present understanding of element cycling will provide the foundation for defining questions and will allow integration of this understanding, from the molecular to global aspects, in the future.

20

SUMMARY OF WORKSHOP-PART 2

Microbial Species

Erko Stackebrandt

DSM-German Collection of Microorganisms
   and Cell Cultures GmbH
Mascheroder Weg 1b
38124 Braunschweig, Germany

SUMMARY OF INTRODUCTION

"A bacterial species is what a taxonomist believes it is, in the broad sense." In the last century, microbiologists followed the system of botanists because early microbiologists were botanists and, of course, adopted the same concepts. During the next 70-80 years, shortcomings of this approach were noted, i.e., descriptions of species based solely on phenotype is of limited value; bacterial species usually do not provide enough information to detect differences; and biological species may not exist among bacteria. At the end of 1970, there were about 45,000 described species of bacteria. The majority of these species were misclassified.

January 1, 1980 was significant for microbiology because the number of microbial species was reduced to those considered to be reliable descriptions of species. Such species were listed by V. B. D. Skerman and co-workers in the "Approved List of Bacterial Names," published in the January 1980 issue of the *International Journal of Systematic Bacteriology* (IJSB). And, all names not on the approved list are considered invalid, with no standing in nomenclature. So, in one day, the number of bacterial species was reduced from *ca.* 45,000 to *ca.* 2,500. Since then, techniques have also been improved. DNA hybridization and chemotaxonomy methods were developed. And, all this information was brought together in a polyphasic approach. Polyphasic taxonomy is an approach proposed by Rita Colwell at a conference in Japan in 1970. Since 1980, the only species validly described are those which are either directly published in the International Journal of Systematic Bacteriology or are listed in validation lists which appear about four times a year in IJSB. There are no other mechanisms for species descriptions not published in the IJSB or listed in the IJSB to be nomenclaturally valid.

There are some points which make species definitions so problematic. The first is that only a small fraction of bacterial species is known and the gaps in the phylogenetic tree, either the RNA tree or hybridization tree, may be due to lack of information. We may not have a tree, we may have a "bush," with very small internodes. And, if it's true that we know

Microbial Diversity in Time and Space, edited by Colwell et al.
Plenum Press, New York, 1996

161

only 0.1% of the species, the rest must provide internodes that are very small, a problem for microbiology as the number of species increases, because there is only a limited set of characters we can investigate by current standards. Discrimination between species will become even more problematic.

A second point to mention is that phenotype alone is not sufficient. One result is creation of "dumping grounds" for bacterial species, such as *Pseudomonas, Bacillus* or *Corynebacterium*, with "messy" genera and "messy" species. Application of molecular methods has made it possible to define the genera better and correct misclassified strains, but always done in combination with phenotype.

Another extreme is that recent history has shown that reducing descriptions to genotype only is not sufficient either. The reason is that there are differences in evolutionary rate. The speed of evolution is different for different bacterial species. For example, 16S RNA similarity or dissimilarity is not enough to establish a number from, for example, 0 to 100 for taxa. Also, due to rather large mutation rates, even cells in one colony may differ from each other. If we are able to measure to the genotypic level or even to the level of the DNA sequence, we will find differences amongst cells in the same colony. Where do we set borderlines? To say each individual nucleotide is an indication of a new species is not acceptable. Freezing an organism at -80°C causes changes in the genotype, for example.

Ecological niches are not known. Often, we are lucky just to isolate an organism, let alone know where it originated, i.e., its actual niche. The degree of genetic exchange often is not known, i.e., how much information is exchanged between closely related organisms or even remotely related organisms. As a compromise, the present genus description or definition is not a concept, but a *description*, based on the polyphasic approach. The first thing to be done is to determine genus, optimally by 16S RNA sequence analysis (or 5S or 23S) or any measure that can measure larger distances. We should avoid creating dumping grounds and place organisms according to phylogeny. 16S RNA is needed to do this, i.e., assign to genus, family, and higher orders (not necessarily at the species level, because it may be an identical sequence, but may be different for the same species). By doing this, we establish an objective database, but the conclusions remain subjective. Thus, a taxonomist sets his or her standard, and it is the taxonomist convincing other taxonomists that their description is optimal for discrimination between neighboring species that prevails. The community of taxonomists accepts this process and the name is then established. If it's not accepted, correction is made. It's a flexible system, with flexibility due to the fact that 70% ± 5-10% DNA hybridization provides room for change.

If a species changes over time, at the phenotypic level, it may differ slightly from other strains of the same species, i.e., it may have drifted sufficiently to comprise a new species. Yet, DNA hybridization above 70% indicates only strain variation. On the other hand, if there is genetic exchange and new genes are introduced or lost, the 70% DNA hybridization won't be affected, because only a few genes are lost, to the extent where differences are so dramatic, a species is formed. In most cases, DNA hybridization is no more than 70% ± 5-10%. The advantage of this present system compared to previous ones, even though not optimal, is that it is flexible, with room for discussion. And, it works.

Question: In response to your comment about acceptance of 70% hybridization. At a workshop a couple of years ago, a circle was drawn with a line through 70% DNA hybridization, because it was concluded not to be valid, that there were too many difficulties with it for use with environmental isolates. Some of the problems had to do with methodology, that it is difficult to use. Other aspects had to do with the fact that it was well tested for medical isolates, but wasn't holding up for a broader range of environmental isolates.

Stackebrandt: Brenner and Johnson were the first to use this approach. They were working mainly with medical organisms, but the 70% threshold value was used because it matches very well with phenotype. Strains that shared more than 70% homology had very similar phenotypic characters. In *Bergey's Manual of Determinatine Bacteriology* or numerical taxonomic studies where more than one strain has been analyzed, there is almost never a total 100% match for each character investigated. There is always some deviation from common properties, but these are strain differences. Another point is that, despite 70% similarity, dramatic differences may occur in the phenotype. On the other hand, 30% hybridization may occur with no differences at all. This is offered by some investigators as a reason to be even more flexible, with perhaps 10% or 20% range. Should we be more flexible? I assume if we find only 30% DNA hybridization, then we should look for other characters than those used, because with that low level of relationship, there must be differences in phenotype. Rather than describing a new species, wait for new data. Another rather extreme example is *Shigella* and *E. coli*, which represent congenera with almost 100% sequence similarity. For medically important organisms, although phylogeny suggests different genera, for the sake of the patient and for ease of diagnosis by the physician, the traditional name is used, even when we know that they actually represent a single species.

Question: If we knew relationships between all prokaryotic organisms on the planet, it would be represented as a bush. Two strains, when compared, if they're the same species, homologies in the "grey" zone may occur. For fungi, the 70% level was rejected because there were so many that are 40, 50, or 60%. It was arbitrary, with the view that for bacteria, it was either that level or less.

Are there many DNA hybridizations for bacteria in the range of 60-40%? Is it an arbitrary number?

Stackebrandt: Yes, its arbitrary. It's a working definition.

Comment: In the case of quite a few well-studied groups, they seem to be units of evolution. For example, *Escherichia coli*, including *Shigella*, seems to be a clearly defined species, genetically distinct from neighboring species, such as *Salmonella*. This is true in quite a few examples. Of course, there are other cases where clear clustering is not observed. This is what occurs in eukaryotes, but the majority of the eukaryotes show clearly-defined species. Humans are a species distinct from other apes, for example, and we have no difficulty in deciding. On the other hand, in many plant groups there are readily recognized species, but in other cases, species are almost as unclear as for the bacteria.

Plants which reproduce largely asexually, e.g., genus *Rubus*, there are many microspecies, which are intergradations, because they reproduce clonally. Taxonomists have trouble deciding whether they are truly separate biological species, in the reproductive sense.

The same will be expected to be true of many bacterial species because bacteria also can reproduce clonally. However, there is a lot of evidence that, within many bacterial species, there is a significant chromosomal gene substitution and exchange which, in all cases where it has been looked for, it's been found. Thus, there does appear to be a mechanism for holding species together in just the same way as the human species is held together by reproduction and recombination, as most eukaryote species are. We should not reject the eukaryotes species definition — the biological species definition — out of hand. We have to determine empirically whether it's true in bacteria or not. We should not say that bacteria don't have species. The question of the level of DNA homology is an interesting one because in eukaryotes, you can't apply a constant level of homology. For example, bird species are differentiated at very much smaller DNA distances than frog species and yet, the biological species concept is applicable to both. We need a different level of cutoff.

The same must be true with the bacteria, which are rather more diverse, compared to each other than birds are from frogs. So, I would expect that 70% might be applicable to enteric bacteria and quite a few other groups of bacteria. But, in some bacterial groups, recombination may occur between genomes that are considerably more diverse. Other groups may have exquisitely sensitive suppression mechanisms that cause them to be isolated even at higher levels. We have to be flexible, as you suggest, but there may well be a biological basis to the apparent groupings that we all hope for and sometimes see.

Stackebrandt: The microbiologist is forced to look at the genetic level. For eukaryotes, it is easier, especially when developing new groups. This may come perhaps with more modern, rapid methods, but at the moment, it's a working basis and it works well.

Comment: In eukaryotes, DNA methods can reveal species which weren't apparent by phenotype. For example, *Drosophila melanogaster* and *Drosophila simulans* look the same, but by using chromosomal or molecular methods, they are completely isolated gene pools. The same may be true of bacteria, which phenotypically look the same, but genetically are different. Parallel phenomena are present in eukaryotes.

Stackebrandt: The range of characters to be included in tests need to be expended. If strains are unrelated at the genetic level, then they will also have differences in phenotype.

Question: Should we think of a bacterial genus as polyphasic?

Stackebrandt: Well, genera are even less well defined. There are no guidelines for defining a genus for bacteria.

Comment: A more highly conserved region of the genome is needed.

Stackebrandt: What I suggest for defining a species is to replace DNA hybridization.

Question: Would 50% DNA hybridization serve for a genus and 70% for species?

Stackebrandt: No.

Simidu: From the point-of-view of an ecologist, the genus exists for bacteria. If we isolate a vast number of bacteria from nature, we are not confident of the existence of species. The problem with present day taxonomy is that it's mostly based on small numbers of "Museum Collections." If someone can isolate many similar bacteria in nature, instead of 10 strains, 100 or 1000 strains, we would not be very confident about clustering at the species level. In nature, species diverge. I don't think there are clear distinctions between one species and another because there are so many intermediate strains. However, we can see more determination at the genus level. For example, photobacteria can be recognized as a separate group. They are distinct in physiological characters, as well as niche. The genus is a real entity, but not species.

Stackebrandt: In the case of *Photobacterium*, it is a subgroup of *Vibrio* and this genus doesn't exist. It falls just within the radiation of the genus *Vibrio*. The alternative is to split the genus *Vibrio* to separate *Photobacterium* as a genus.

Simidu: The genus *Vibrio* should be split. In deep seawater, many different strains of *Photobacterium* can be found, even as dominant as marine *vibrios*.

Stackebrandt: Luminescence is a convenient character. In many cases, it's not so obvious that a single character unifies a genus. Chemotaxonomy then becomes necessary.

Simidu: We came to this conclusion by analyzing 16S ribosomal RNA of many natural isolates from deep water.

Comment: It's true that, in many cases, those groups that have been defined on the basis of a conspicuous property like luminescence, type of photosynthesis, or root nodulation turned out to be phylogenetically heterogenous. Maybe it is fortunate, in this case, that luminescence is confined to one branch of phylogeny, which is not true for many other physiological properties.

Stackebrandt: The sequence can be retrieved from a clone library for organisms targeted by probes for direct detection in the environment. The organism, perhaps by shape, can be observed in the environment. A third case is when the organism has been identified, perhaps as a symbiont and can be characterized by morphology and amplify the genome with random primers and GC content determined. If flagella are present, more information is then available. And, for this case, there is a proposal, in IJSB by Schleifer and Murray to assess this uncertain taxonomic status. The nomenclature code for bacteria does not require culture. One can describe a species even if it cannot be grown in pure culture. *Planctomyces* is an example for which a pure culture has not been grown. Nevertheless, taxonomists can provide information sufficient to distinguish it from its neighboring taxa. The code allows for this, without hybridization

Question: These are not available in pure culture? Is it necessary to have it in culture?

Stackebrandt: It must be a living culture, detectable in a specimen, e.g., insect.

Question: DNA sequences detected in the environment would not be eligible for description as a species?

Stackebrandt: No. That may be the next step. Taxonomists are conservative. It will take time to go from a species in culture to a sequence, alone, as a valid species.

Question: What is the next stage beyond DNA hybridization? It should be important to find substitutions, another gene?

Stackebrandt: It might be a gene that can be easily amplified and is less conserved than 16S or ribosomal RNAs. Unfortunately, genes have a degenerative code. It's not easy to develop conserved primers. We have tried heat shock genes and failed. There are no conserved primers for heat shock protein genes. All the others, such as ATPase and translation factors are too conserved. But, there must be something which is reproducible and less conserved. We want to cover the upper range of the relationship to replace DNA hybridization. Molecular taxonomists are asked to participate in a search for this gene.

Comment: 16S ribosomal RNA information itself has limitations in delineating species. Can we generate differences that correlate with phenotypic differences? We have found some inconsistencies between phenotype and genotype. There is evidence for transfer of genes encoding functionally important proteins. Phototrophic purple bacteria can be classified into three major groups: alpha; beta; and gamma groups within the bacteria, but the phylogenic tree of the phototrophic bacteria is based on the purple bacterial gene coding for a photosynthetic gene. It's different from the tree based on ribosomal RNA sequences. Some members of alpha phototrophic bacteria are more closely related to the beta and gamma clusters than to each other within the alpha subgroup. This suggests the possibility of gene transfer of the photosynthetic gene. This is the case of *Pseudomonas* and their nonphototrophic relatives, as shown by Dr. Yen.

Stackebrandt: We have not observed 70% hybridization between physiologically and phenotypically absolutely unrelated organisms. I don't know of any case.

Question: Even if the hybridization is greater than 70%?

Stackebrandt: There are always exceptions to a rule. If there is such a case, one would probably split these organisms into different genera, as one alternative. Or, if one is a mutant of a phototrophic genus, it would be extraordinary to lose the whole photosynthetic apparatus. These occur with artificially-induced mutants.

But, your question is a good example that one should also avoid using sequences that are involved in horizontal gene transfer. The problem with DNA hybridization is that sequence analysis was developed later. Everyone today could live with species delineation at 100% RNA homology. It's reproducible, easy, and would replace tedious hybridizations, but it's difficult to change the system.

GENERAL DISCUSSION

Cohen: We have had a very productive three days, and it is time to discuss where we go, what are the important steps to take to understand the function of mixed communities of microorganisms, and their inter-relationship with their environment. First, let's discuss the structure of the community and its heterogeneity. This links with the possible heterogeneity of marine snow particles, for example, which may or may not constitute important microenvironments within a seemingly homogenous, often oligotrophic medium of central tropical seas. With respect to community structure, it's a beginning, since we don't have the organisms in culture. We can say that they are the exciting 90% of uncultured organisms. The main breakthrough would be to actually isolate the organisms, but we have to develop new concepts to enrich for and isolate that part of the microbial world not in culture.

We should address how to go about developing functional probes. Probes that would tell us what these communities are doing *in situ*. It seems easy, but it's not. It's not trivial to develop a probe for a pure culture in the laboratory and take this probe to a mixed culture to an heterogeneous world, and get results that are meaningful and well controlled. Functional probes and assessing activity are important.

Discussion: A potential way to think about the concept of cultivation is polyphasic ecology. What I mean by that is any one of these approaches. If we just study phylogenetic diversity, we don't know what functions the organisms are carrying out. If we study only biogeochemistry, we won't know what the organisms are that are responsible for the bioconversions. If we study only the genes that are being expressed, even if we identify the organisms they're coming from, we don't know if the actual activity is being expressed. A good example is nitrogenase. In some cyanobacteria, there are basically post-transcriptional modifications which are largely part of the regulatory mechanism for nitrogenase enzyme expression. There are many levels of regulation beyond mRNA transcription. Any one of these approaches has weak points. The links will eventually provide the whole picture.

Many cultivation techniques do work and work well, but are they necessarily going to work well for all organisms? Is the way in which we, as microbiologists, think about cultivation skewed? When we think about cultivation, what do we mean? In a general sense, it is growing microorganisms to high densities that we can measure by optical density in a test tube, *ca.* 10^7 or higher, or what we can see as colonies on a petri dish. Very high concentrations. For many habitats, unnaturally high concentrations. But, that's what we mean by "cultivation." That's how we detect growth, either by eye on media or in liquid, or with a spectrophotometer. Dr. Button, at the University of Alaska, made dilution cultures of seawater, starting with 10^5 cells per ml natural seawater, diluted with sterilized natural seawater (basically an MPN). What he found was that when he did that, he could get the microorganisms representing the majority to grow in extinction dilution. A large proportion of that population was able to grow. He was able to detect growth by measuring it with a flow cytometer. These cells peaked after exponential phase at about 10^4 per ml, maybe a little higher, but that was how concentrated they grew to. The cells were growing in test tubes. But, we would say that they are not cultivatable because they didn't reach densities of 10^7 or 10^8. We need to think about what we mean by "cultivation" and whether it is a loaded term. To get some of these hard to cultivate organisms, we're going to have to think beyond colonies or densities of 10^7 per ml.

Comment: The polyphasic approach: big programs can be as important, as individual studies. There are many here who are very interested in this concept of who's out there and what are they doing? What we should do is call for a human genome type of project with 16S, cultivation, microprobes, and many of the other new technologies and start with one system. A microbial system, perhaps a mat or snow and develop a multi-year, a multi-investigator project to study the structure and function of a system, with a scientific hypothesis driving the project.

With respect to cultivation, people who work with fastidious phytoplankton have taken the initiative to go to tremendous lengths to succeed in achieving culture of phytoplankton species, e.g., *Prochlorococcus*, in culture. They take an extreme approach and the cells don't achieve very high concentrations. They count with a flow cytometer. People who work with heterotrophic bacteria traditionally have not been as fastidious. Certainly people who are working with anaerobes and some of the sulfur-oxidizing groups of bacteria that are notoriously hard to grow have taken very unique kinds of approaches. Those who work with heterotrophic bacteria have made the excuse that the bacteria are not cultivatable instead of doing the hard work to get them into culture.

Stackebrandt: Concerning the polyphasic approach, what we all would like to see is development of a network of strategies. We can then select a part of the strategy. Concerning a site to concentrate on taxa for biodiversity inventory at the larger scale, in Europe, the E.C. organization deals with biodiversity. People could not decide on sites to investigate because the people were from different disciplines: marine, freshwater, soil and they would rather develop methods than all work on one model. This is something we have to do, i.e., develop a network of strategies.

Comment: To prioritize, for the first five years, a tropical or freshwater lake. Maybe the next five years, it might be an arctic or boreal region. The real value is to share the same sample at the same time. It's amazing the similarities that people are finding, like the *Archaea*.

Stackebrandt: Different laboratories using the same method should be involved to test the validity of certain techniques. The international working group for *Mycobacterium* taxonomy has worked together for 20 years, always involving different laboratories doing similar studies, to evaluate methods for *Mycobacterium* taxonomy. They use coded samples and compare methods.

Comment: It would be highly desirable if we could agree on an intensive study of two types of sites, e.g., ocean and terrestrial. It's important to have two contrasting sites because then the principles learned from one or the other are more robust. We can use the argument that those two environments are very different and the advantages and disadvantages complement each other. We can compare ecological principles.

Comment: A pristine or a typical base site can serve as reference for impacted sites.

Comment: This is more attractive than a total inventory. At least for some of us that is a bit foreign because we're not sure that this really addresses the major biological problems that we would like to address. The scheme that has just been presented is attractive.

Comment: The Global Change Program and JGOFS are big ocean programs. Most people are interested in global oceanic carbon cycles, but to study this, microbial groups in surface waters are important. To do this, you need to investigate micro-scale phenomena, as with marine snow. Even in terrestrial areas, we can define some very important processes.

Simidu: ICOME has been making a number of efforts to promote microbial diversity, although funds are limited. One of the important issues is how we can persuade policymakers, citizens, and scientists outside of microbiology about the importance of microbial diversity. If we simply say that microbial diversity is important, they'll say that's natural, but we must select a particular problem. We must select key phenomena, key processes, key geographical sites, key areas that might interest policymakers and biologists, chemists, and other scientists who don't know very much about microorganisms.

Comment: Just about every biogeochemical process on this planet is mediated in some way by microorganisms. It really doesn't take too much imagination; there's all sorts of problems involving methane, nitrogen, nitric oxides, CO_2, bioremediation, oil spills, the list is endless.

Comment: One example of why people are interested in microbes in terrestrial systems is that we don't know the extent of decomposition of material in soil. It's a big black box. We can't do estimations of the total budget of carbon dioxide in the atmosphere or the oceans.

Comment: At one level, we need to decide to work together and what project to work on. If it is practical, it is legitimate. Assuming we agree on a biotope to concentrate on, do we agree what is important to be done? A multiple approach to understanding a system is obvious. Can we think of what is our present limitation in the understanding of a system, with respect to not having sufficient technology or not being able to get to the scale at which we wish to understand in space or time? We should discuss the scientific limitation in understanding biodiversity at the ecosystem level.

Comment: The over-all objective would be to focus on linking process and populations, with in-depth understanding at several levels, molecular and community, and also up to larger scales.

Comment: Would it be appropriate to have a series of workshops with experts in individual fields presenting their methods for culturing microorganisms? A workshop on novel isolation procedures, on modern cloning/sequencing, on new physiological methods?

Comment: Culturing might be particularly timely because people might have forgotten about that, with such interest now in the molecular. It might reemphasize its importance and also offer ideas about how to do things beyond standard methods.

Comment: We're always thinking of isolating pure cultures, but the idea of preparing mixed cultures where one takes one component which is already isolated as a pure culture, but provides something that the other component needs, can be informative.

Stackebrandt: "Microbial Diversity in Time and Space" is a wonderful title for a global program.

Simidu: In closing the discussion, I'd like to express thanks to all the participants, particularly to our colleagues who traveled a long distance. We had a very pleasant meeting and very fruitful discussions. We'd like to continue our efforts, to develop a network, and research projects that can contribute to the microbial diversity program in Japan and in every other country. I'm convinced that some day politicians, as well as biologists will fully appreciate the importance of microbial diversity, and they will provide substantial support for this effort.

INDEX